WE ARE AS
GODS

Some thoughts on *Abundance*, fourteen years later...

Back in 2012 when we published *Abundance: The Future is Better Than You Think*, the concept of abundance was still a distant prediction. Now it's become a rallying cry for global technology leaders.

"We're heading into a future where technology will unlock abundance in education, healthcare, and opportunity for everyone."
—Jensen Huang, CEO of Nvidia

"Artificial intelligence promises a future of unparalleled abundance."
—Vinod Khosla, founder of Khosla Ventures

"There will be an age of abundance. No shortage of goods and services. Everyone will have everything they want. The cost of goods and services will trend to zero.... A future where there is no poverty."
—Elon Musk, CEO of xAI, Tesla, SpaceX

"As we enter the AI era, we predict the benefits to consumers will be as profound as the technology is magical.... In a word, we predict an era of abundance—consumers' lives will be enriched through new channels for creativity and self-expression, new paths to self-discovery and belonging, and new ways to do the most meaningful work of their lives."
—Marc Andreessen, cofounder of a16z

"I think we see a path now where the world gets much more abundant and much better every year.... With superintelligence, we can do anything else ... massively increase abundance and prosperity."
—Sam Altman, CEO of OpenAI

"Some of the things we get to look forward to ... near limitless intelligence and abundant energy ... could fix climate, establish a space colony ... an age of abundance."
—Garry Tan, CEO of Y Combinator

"Technology is not just a tool. It can be a force for good, a force for change, a force for a more abundant future."
—Tim Cook, CEO of Apple

"The abundance of information, if deployed correctly, should lead to an abundance of democracy."
—Eric Schmidt, former CEO of Google

"I see a world of abundance ahead, where science and technology are used to solve the world's greatest inequities."
—Bill Gates, cofounder of Microsoft

What the world's greatest technologists and thought leaders are saying about *We Are as Gods*...

"*We Are as Gods* is the critical reading for the coming Artificial General Intelligence and Singularity. Peter Diamandis and Steven Kotler show that while exponential technologies deliver the *capability* for radical abundance, the real challenge lies in upgrading our *consciousness* to match our accelerating power. This is more than a survival guide—it's a manual for optimizing our destiny, connecting the speed of technology to the unlimited potential of the human mind."

—Ray Kurzweil, inventor, futurist, cofounder of Singularity University, and author of *The Singularity Is Nearer*

"Diamandis and Kotler's bestseller *Abundance* helped shift the global conversation from fear to possibility. Now, *We Are as Gods* reveals that the forces they predicted—AI, clean energy, digital biology—are scaling at a pace few could imagine. This book argues persuasively that the Abundance era has arrived and challenges leaders to use these capabilities responsibly and ambitiously. A timely and important follow-on to a landmark work."

—Eric Schmidt, PhD, former CEO of Google, CEO of Relativity Space

"*We Are as Gods* is a guidebook for the age when human intelligence becomes networked with machine intelligence. Peter Diamandis and Steven Kotler show us how exponential technologies—AI, robotics, synthetic biology, and beyond—don't just give us new tools, they give us new leverage to solve problems at societal scale. This book is essential for any entrepreneur or leader who wants to build a future defined not by limits, but by compounding possibility."

—Reid Hoffman, cofounder of LinkedIn, partner at Greylock, and coauthor of *Superagency: What Could Possibly Go Right with Our AI Future*

"Change is accelerating at an exponential pace. Diamandis and Kotler show you how to harness AI and converging technologies to create abundance, purpose, and true agency in your life. If you're ready to stop feeling overwhelmed and start taking control of your future, *We Are as Gods* is your guide for mastering an abundant future—and maximizing your impact now."

—Tony Robbins, the world's #1 life and business strategist and four-time *New York Times* bestselling author

"In *We Are as Gods*, Peter and Steven capture the same spirit that drives every XPRIZE: the belief that humanity's greatest challenges are solvable. This book is a powerful reminder that exponential technologies are dramatically expanding our reach—on earth and beyond. It's a must-read that invites us to think boldly, act courageously, and design a future worthy of our highest aspirations."

—Dr. Anousheh Ansari, CEO, XPRIZE Foundation, and first private female astronaut on the International Space Station

"This book redefines what progress should mean—not just more, but better. Diamandis and Kotler push us to build a future of shared prosperity, where abundance is measured by justice, creativity, and human potential. *We Are as Gods* is the conversation our times demand."

—Van Jones, CNN host, founder of DreamMachine.org, and *New York Times* bestselling author of *Rebuild the Dream*

"An extraordinary blueprint for hope. *We Are as Gods*—both exhilarating and grounding—shows how exponential technologies can serve our deepest human values. This book acts as an invitation and a practical guide at the time when humanity needs it most. It turns optimism into a discipline and a mindset that empowers us to build a hopeful and compelling abundant future. Diamandis and Kotler once again offer us all reason for optimism."

—Mo Gawdat, former chief business officer of Google X and bestselling author of *Scary Smart: The Future of Artificial Intelligence and How You Can Save Our World*

ALSO IN THE EXPONENTIAL MINDSET SERIES

The Future Is Faster Than You Think: How Converging Technologies Are Transforming Business, Industries, and Our Lives

BOLD: How to Go Big, Create Wealth, and Impact the World

Abundance: The Future Is Better Than You Think

WE ARE AS GODS

A Survival Guide for the Age of Abundance

PETER H. DIAMANDIS • STEVEN KOTLER

Atlantic Books
London

First published in the United States in 2026 by Simon & Schuster,
1230 Avenue of the Americas, New York, NY 10020.

First published in hardback in Great Britain in 2026 by
Atlantic Books, an imprint of Atlantic Books Ltd.

Copyright © PHD Ventures, Inc. and Steven Kotler, 2026

The moral right of PHD Ventures, Inc. and Steven Kotler to be
identified as the authors of this work has been asserted by them in
accordance with the Copyright, Designs and Patents Act of 1988.

All rights reserved. No part of this publication may be reproduced, stored in
a retrieval system, or transmitted in any form or by any means, electronic,
mechanical, photocopying, recording, or otherwise, without the prior permission
of both the copyright owner and the above publisher of this book.

No part of this book may be used in any manner in the learning,
training or development of generative artificial intelligence technologies
(including but not limited to machine learning models and large language
models (LLMs)), whether by data scraping, data mining or use in any
way to create or form a part of data sets or in any other way.

Every effort has been made to trace or contact all copyright holders.
The publishers will be pleased to make good any omissions or rectify
any mistakes brought to their attention at the earliest opportunity.

10 9 8 7 6 5 4 3 2 1

A CIP catalogue record for this book is available from the British Library.

Hardback ISBN: 978 1 80546 489 1
Trade Paperback ISBN: 978 1 80546 490 7
E-book ISBN: 978 1 80546 491 4

Printed and bound by CPI (UK) Ltd, Croydon CR0 4YY

Atlantic Books
An imprint of Atlantic Books Ltd
Ormond House
26–27 Boswell Street
London
WC1N 3JZ

www.atlantic-books.co.uk

Product safety EU representative: Authorised Rep Compliance Ltd., Ground Floor,
71 Lower Baggot Street, Dublin, D02 P593, Ireland. www.arccompliance.com

Peter's Dedication

To my family, Jet, Dax, and Kristen—in a world where we are becoming as gods, you remind me that our humanity is not something to transcend but something to treasure. This book is for you and for the world you'll help create. May you wield these godlike powers with humility, curiosity, and an unshakable commitment to human flourishing.

Steven's Dedication

To Joy, for still holding my hand on this adventure.

CONTENTS

PART 1: **WARP SPEED**

Chapter 1: **Theogony**	3
Chapter 2: **An Abundance of Abundance**	23
Chapter 3: **Data-Driven Optimism**	41

PART 2: **EVERYTHING, EVERYWHERE, ALL THE TIME**

Chapter 4: **One Billion Times Smarter**	59
Chapter 5: **Surfing the Tsunami**	79
Chapter 6: **The Dark Side of Abundance**	101

PART 3: **THE AGE OF ABUNDANCE: A SURVIVAL GUIDE**

Chapter 7: **Mind 2.0**	129
Chapter 8: **The Androids Are Us**	155
Chapter 9: **The Paradise Paradox**	177
Coda	201

Appendixes

    A: Proof of Abundance Charts   203

    B: Charts Showing the Dark Side of Abundance   233

    C: Exponential Technology Driving Increased Abundance   243

Acknowledgments   261

Notes   263

PART 1

WARP SPEED

This is magic. Sure—but not necessarily fantasy.
—THOMAS PYNCHON

CHAPTER ONE

Theogony*

A Tall Order

"We are as gods and we might as well get good at it," wrote *Whole Earth Catalog* founder Stewart Brand in 1968. He was talking about technology. It was the start of the space age and the dawn of the computer era. The excitement was palpable. But it was a tall order.

Consider a few of the godlike miracles we'd have to master, like *creatio ex nihilo*, the creation of something from nothing. Now, there's a divine attribute rare in the universe of comparative mythology. You want to turn formless nothing into something? Anything? That feat is reserved. Only supreme deities need apply.

Yahweh, the God of the Hebrew Bible, pulled it off. He dragged light, land, and life out of the void. He qualifies. As does Brahma, who turned pure potential into a cosmic egg. Io in the Māori tradition, Atum in Egyptian cosmology, Pangu in Chinese lore, all pulled the rabbit of creation out of the hat of, well, nothing. Not even a hat.

Here's another divine superpower: omniscience. The ability to

---

* Theogony (*noun*): a work describing the origins and genealogy of the gods, traditionally a narrative or an epic poem.

know all things. A third: omnipresence. The ability to exist everywhere at once. Without question, the old gods could razzle-dazzle.

Praecognitio. The ability to foretell the future. The always-popular *praesentia invisibilis*, which is the ability to be invisible—technically "presence without being seen," more technically, "invisible presence."

Or shape-shifting, *mutatio formae* in the Latin, *metamorphosis* in Greek. This is one of the rarest supernatural talents. Kali could pull it off, as could Proteus. And then there's Loki, the enigmatic trickster in Norse mythology, who once transformed himself into a beautiful mare, was impregnated by a fierce stallion, and gave birth to Sleipnir, the eight-legged horse ridden by Odin.

In fact, if you thumb through the Old Testament counting superpowers, the standard total is eighty-three miracles that fall into ten major categories. Mind you, to create this list, we've ignored the global canon and confined our search to a single book. Still, here's the breakdown:

Creation Miracles: 1
Provision Miracles: 10
Nature Miracles: 16
Healing Miracles: 7
Resurrection Miracles: 3
Judgment Miracles: 15
Protection Miracles: 12
Prophetic Miracles: 9
Communication Miracles: 5
Victory-in-Battle Miracles: 5
Total Old Testament Miracles: 83

That's a lot of miracles. Plus, Stewart Brand made this proclamation in 1968—in the early days of godlike technology. Back then, mainframe computers the size of oil tankers were the rule, the microchip had just been born, and color television remained a neat trick. Yet we weren't nowhere.

In 1968, push-button phones were suddenly a thing. So was spaceflight. Of course, Neil Armstrong's small step was still a year away, but NASA did manage to orbit the moon with a man in the can—technically, three men in a big can, including Jim Lovell, who later became famous for being the commander of the ill-fated Apollo 13 mission. At the time, he was only a navigator. Still, imagine being the first person to chart a course around the dark side of the moon. . . .

So yeah, in 1968, when Brand's famous dictate first appeared, we were not quite gods. We were gods in training. But what nobody quite expected—we were very fast learners.

Theurgicon*

"I was blind, but now I see," reads the New Testament. The statement describes one of the more famed miracles in Scripture: Jesus's restoration of sight to the blind. Technically, a miracle of healing. But in this case, the small print is important.

The once-blind man healed by Jesus makes his claim while being interrogated by the Pharisees, the spiritual leaders of the time. Essentially, it is a statement given in religious court, under oath, as a testament to Jesus's supernatural abilities and de facto proof, in the mathematical sense of that term, of his divine origin. Like the Resurrection, the raising of the dead, and the parting of the Red Sea, the restoration of sight to the blind has become a baseline metric for divine power, which makes it a useful measuring stick for progress here in the early twenty-first century. In the calculus of Stewart Brand, the restoration of sight to the blind is a way to judge just how good we humans have gotten at behaving like gods.

This brings us to Max Hodak, who is founder of the Science Corporation, a company that is in the miracle business—using microscopic

---

* Theurgicon (*noun*): a divine academy, a place where gods go to learn how to wield their superpowers.

retinal implants and infrared optics to restore sight to the blind. A biomedical engineer by training, Hodak's primary focus is brain-machine interfaces, which is to say, the miracle business is business as usual for Max Hodak.

Before starting Science, Hodak co-founded and served as president of Elon Musk's brain-computer interface company Neuralink. Neuralink's mission is to build brain implants that let people with paralysis control devices with their thoughts. Technically, Neuralink is a triple miracle because their implant doesn't just heal paralysis, it also facilitates telepathy and telekinesis. Still, if we factor impact into our assessment, the PRIMA retinal prosthetic, as the Science Corporation's implant is known, wins the battle for scale.

Twenty million people suffer from spinal cord injuries. It's both the leading cause of paralysis and the trauma that Neuralink's device is meant to heal. The PRIMA implant, meanwhile, combats age-related macular degeneration, which is the leading cause of blindness in the world and a disease that affects over 170 million people. In other words, curing macular degeneration really is a miracle of biblical proportion.

Hodak's cure for blindness is a two-millimeter photovoltaic microchip containing nearly four hundred light-powered pixels that replace the retina's normal photoreceptors—the ones lost to macular degeneration. A camera mounted on a pair of glasses captures visual information, which is projected as a pattern of near-infrared light onto the chip's photodiodes. The chip works like a solar panel, converting infrared light into electrical signals that stimulate the surviving inner retinal neurons. These neurons transmit the signals along the optic nerve to the visual cortex, where the brain constructs them into images, mimicking the process of natural sight.

The surgery required to implant the microchip takes about eighty minutes. The results, pardon the pun, are eye-opening. In the United States, normal vision is 20/20, while 20/200 or worse is the threshold for legal blindness. In a trial of the implant run in the United Kingdom and Europe, thirty-two subjects started the study with 20/450 vision—legally blind by US standards. After receiving the device and measuring visual acuity with the standard eye chart, all thirty-two people improved their performance by twenty-three letters, which is five lines down the chart from where they started. On average, postsurgery, eyesight improved to 20/160—the difference between seeing darkness and seeing faces (aka, a miracle). Because of the globe's aging population, by 2040, there will be nearly three million people in the world blinded by macular degeneration. Restoring vision for three million people—that's the scale of miracle we're describing.

Yet this raises a crucial question: What happened between 1968, when Stewart Brand first made his pronouncement, and today, when we can have a science-based discussion about humankind's newfound ability to perform miracles of biblical proportions?

The answer is exponential technology.

Any technology that doubles in performance while dropping in price on a regular basis is an exponential technology. Moore's law is the standard example. In 1965, three years before Brand's "we are as gods" pronouncement, Gordon Moore caught sight of a similar trend. He noticed the number of integrated circuits on a computer chip had been doubling every eighteen months while the cost of the chip remained the same. This is a classic example of an exponential doubling, a so-called price-performance curve, where costs stay constant and performance improves on a regular basis.

But really, what Gordon Moore noticed was the emergence of a piggyback ride. Once a technology becomes digital and can be translated into the ones and zeroes of computer code, it jumps onto the back of Moore's law and begins accelerating exponentially. Every doubling in computing power feeds a doubling in that technology's capability—which in turn drives further improvements in computing. Progress compounds upon progress, producing the runaway effect we now call exponential acceleration.

In 2012, we introduced the concept of exponential technology in our first book, *Abundance: The Future Is Better Than You Think*. In it, we held our ground against naysayers and pessimists and made bold, evidence-based predictions about technology's ability to improve standards of living and where the future was heading. We examined ten technologies accelerating on exponential growth curves—computers, sensors, networks, AI, robotics, 3D printing, augmented and virtual reality, biotechnology, and blockchain—that would soon give humanity the ability to meet the basic needs of every person on the planet. In short, in terms of godlike powers, *Abundance* was a data-driven prophecy about the miracle of provision.

In the years since that book's release—as will be explored in greater detail throughout these chapters—exponential technology has made good on this promise. The world has witnessed measurable increases in a host of critical abundance-related metrics: per capita income, access to food, energy, communications, education, healthcare—the list goes on. Truthfully, there's no end in sight. Simply stated, if standards of living are your metric, more people are living better lives than ever before in history.

The rise in computational power that is driving this progress—what was the quaint, yearly doubling known as Moore's law when we wrote *Abundance*—exploded into tenfold annual growth between 2012 and 2022. Today, riding the triple engine of GPU acceleration, exponential data growth, and ongoing breakthroughs in generative AI, it's surging toward one-hundred-fold gains—godlike power indeed.

The Science Corporation and the PRIMA retinal prosthetic are where these surges have led. The implant is the result of exponential acceleration in a half-dozen fields and, more specifically, the *acceleration of acceleration* produced by converging exponentials. When waves of development for different accelerating exponentials collide, progress stops compounding—it detonates. These intersecting waves stack on top of one another, accelerating acceleration by doubling in power and size and producing a whole much greater than the sum of its parts.

These same forces are now transforming entire industries—especially at the intersection of AI and robotics, two of the central technologies fueling today's age of abundance. In 2012 when we wrote *Abundance*, we predicted a future for these technologies that included everything from autonomous cars and flying cars to autonomous robots and delivery drones. It really was a prediction of transportation abundance—the safer, cleaner, and cheaper movement of goods and people than ever before in history.

Today, our prediction is reality. Thanks to the power of convergence, there are over thirty autonomous car companies, and nearly every major retailer has robots running their warehouses. Flying car companies are operational in the Middle East and Asia, and companies like Zipline are making thousands of drone deliveries every day, transporting lifesaving medicines, and saving tens of thousands of lives in the process. In a little more than a decade, we have gone from hard-to-believe stories about flying cars and robo-maids to commercial operations of electric vertical takeoff and landing vehicles (eVTOL), and internet videos of humanoid robots folding clothing, serving drinks, and holding yoga poses (google: Optimus). And transportation is only the beginning.

The same converging exponentials powering these new industries are also helping Hodak cure blindness. The PRIMA implant sits at the intersection of four accelerating technologies: computing, artificial intelligence, nano-fab electronics, and material science. Individually, each of these technologies is revolutionary. Together, fueled by convergence, they're a paradigm shift—one that requires miracle metaphors to explain.

Convergence is why it's no longer hyperbole to say that our ancestors would view us as gods. The blind can now see. The paralyzed can walk. While no one has yet multiplied loaves and fishes to feed the hungry, we can now grow fish from stem cells to accomplish the same miracle of provision. Certainly, the terminology has changed. The grandiosity of "omnipresence" and "omniscience" has been replaced by the prosaic "Zoom" and "Google"—but the underlying superpowers are the same.

Here's the catch. It's not just Max Hodak. It's all of us. These divine powers are everywhere and everywhen. They're now in your pocket. You carry miracle tech in your jeans and handbag. You can summon a chariot of the gods disguised as an Uber with a finger tap and conjure up a feast via Uber Eats with another. You have answers to nearly every question in seconds. Translation, navigation, simulation—it's all on demand.

So, here's the next question: If we are literally walking the earth in an age of miracles, how come we don't feel divine?

Structure Mapping

If the divine feels distant, it's because our brains weren't built to process miracles at scale. Without grounds for comparison, novelty overloads the system. The unfamiliar floods perception and the wonders of the world register as fear and confusion. To make sense of an age of miracles, we need a way to translate the extraordinary into the understandable—a mapping system for the mind. That's where the work of Northwestern University cognitive scientist Dedre Gentner comes into our story.

In the 1980s, Gentner ran a series of experiments investigating how humans come to understand unfamiliar concepts. She asked people questions like, "How is a solar system like an atom?" or "How is a battery like a reservoir?" and examined how they reached their answers.

In the case of comparing solar systems to atoms, most people explained the relationship by saying something like: "Electrons orbit the nucleus like planets orbit the sun." In other words, when people try to understand atoms, they used space analogies—solar systems, orbits, gravitation. But Gentner's discovered something deeper about how the mind works, a process she termed *structure-mapping*.

In order to understand the unfamiliar, humans don't just make surface-level comparisons. We map deep relational similarities between domains. In Gentner's experiment, people were not saying atoms *look like* solar systems. They were saying: The structure of orbiting bodies under invisible forces in one domain helps explain the other.

This insight about analogy—that we understand new things by mapping them onto known structures—has become foundational to how we think about thinking. Analogy is now viewed as cognitive infrastructure for the mind. If you agree with philosopher Douglas Hofstadter, it's the root of intelligence and creativity. It's how we learn to understand the new, the unfamiliar, even the incomprehensible.

In the twentieth century, the most transformative analogies reshaped entire sciences. Neuroscience, for instance, took a quantum leap once researchers began comparing the brain to a computer. Suddenly, cognition could be modeled in terms of information processing—inputs, outputs, memory, feedback loops. That analogy gave rise to computational neuroscience, artificial intelligence, and brain-computer interfaces (BCI) like Hodak's implant. The same principle applies at every scale: When we describe the internet as a "worldwide web," or genes as "code," or the universe as a "network," we're not just explaining—we're expanding what's possible to imagine.

Analogy is the brain's way of compressing novelty into familiarity. It's how we make sense of the world when it starts changing faster than we can keep up. But this ancient cognitive tool—structure mapping—fails in the face of modern technology. Our comparison machinery runs out of easy comparisons. Godlike powers in our pockets? What's the analogy here?

Without grounds for comparison, we can't parse the world. The result is cognitive vertigo—the sense that the world is moving faster than we can make sense of it. And if we can't reason our way through change, we myth our way into progress. When analogies fail, humans start hunting deeper patterns—meta-analogies of a sort—or what the Swiss psychiatrist Carl Jung called *archetypes*.

In Jung's definition, archetypes are universal, inherited patterns of thought, image, and idea that are embedded in humanity's collective unconscious. As a result, archetypes are primal symbols that evoke powerful reactions across cultures and generations. The Hero, the Shadow, the Great Mother, the Wise Old Man, these figures materialize in our myths, manifest in our dreams, and take form in our art—shape-shifting the human narrative by influencing our perception of self, other, and the world.

In the early twenty-first century, we find ourselves at a loss for easy analogies yet awash in Jungian archetypes. Between 1968, when Brand made his pronouncement about our godlike potential, and today, when that potential is starting to be realized, one way to track the impact of technological acceleration on our psychology—call it the failure of analogy—is to track the rise of *archetypal media*, adding up the gods, goddesses, superheroes, and supervillains populating our screens to see what those numbers reveal about the modern mind.

If you begin in 1968, the next decade saw one notable cinematic release, *Superman*, a.k.a. the Hero, while television produced *Wonder Woman*, a.k.a. the Heroine. The 1980s saw the first step-function in amplification, with ten superhero films, including *Batman* and *Superman II*, and six TV shows, ranging from *Transformers* to *The Incredible Hulk*. Things doubled again in the 1990s, with twenty major cinematic releases and nearly that many TV shows. But between 2000 and 2010, the numbers triple: sixty films and thirty television series.

Jung would argue that this surge in archetypal media is an unconscious response to the psychic destabilization brought on by the radical acceleration in human potential. With each technological leap forward, there's a parallel need for new symbols and myths to anchor our understanding of our growing power. Archetypes provide narrative coherence and moral clarity. When humans face destabilizing change, archetypal figures emerge to help us integrate new abilities, often reflecting society's greatest hopes and fears. As Spider-Man says: "With great power comes great responsibility."

We live in a world of abundant archetypes because we live in a world of abundant miracles. They're the psychological byproduct of technological acceleration and its startling ability to help us solve intractable problems. In fact, if you chart major techno-sociocultural trends and their impact on the rate of change in the world, you find the pace of change accelerating 233 percent faster than it did in 2010. The results: Humans have superpowers.

Put differently, if we measure exponential progress against the Old

Testament's ten miracle categories, the results are—well, see for yourself.

But be prepared. This is a very long list.

Creation Miracles

- Synthetic biology and genetic engineering create new forms of life or modify existing ones.
- 3D printing brings matter into being layer by layer, building something from nothing.
- Generative AI creates virtual worlds populated by self-directed agents that are capable of forming new economies, religions, and societies.

Provision Miracles

- Vertical farming, desalination, and lab-grown meat provide a bounty of food and water with minimal land, waste, or suffering.
- Genetically modified crops and artificial photosynthesis turn sunlight into food and fuel.
- Solar-powered water purifiers, ovens, and aquaponic systems provide clean water, cooking, and sustainable protein even in deserts or disaster zones.
- Drones deliver meals and medicine where supply chains fail.

Nature Miracles

- Geoengineering, cloud seeding, and precision irrigation lets us steer weather and prevent drought.
- Disaster-prediction networks now forecast hurricanes, earthquakes,

and tornadoes with astonishing accuracy, while lightning-control lasers and wildfire drones prevent small sparks from becoming infernos.
- Carbon capture reduces atmospheric CO_2, while reforestation and anti-desertification efforts—tree-planting drones, high-efficiency crops, and water-saving irrigation—are feeding people and turning barren land green again.
- Aquatic robots clean pollutants from rivers and oceans, restoring ecosystems.

Healing Miracles

- Gene therapy and CRISPR cure disease at a genetic level.
- Stem cell therapy and tissue engineering repair what disease and injury destroy.
- Epigenetic reprogramming can reverse ocular degeneration in animals, and soon will be able to do the same in humans.
- Telemedicine enables the remote diagnosis and treatment of disease.
- Advanced diagnostics, such as imaging, genomics, and metabolic profiling, allow physicians to detect and intervene in seven out of the top ten causes of death, often before symptoms appear.
- AI-powered drug discovery accelerates cures for complex diseases, such as cancer and rare genetic disorders.

Resurrection Miracles

- Cryonics preserves the dead in the hope of future revival.
- Organogenesis uses stem cells to create new organs.
- Organ profusion and preservation technology allow donated organs to survive for days instead of hours, saving lives in the process.
- Cardiopulmonary resuscitation (CPR) and drone-delivered defibrillators can revive people in cardiac arrest, a modern form of resurrection.

Judgment Miracles

- AI-powered surveillance—facial recognition, drones, and predictive policing—monitors, forecasts, and prevents crime.
- Autonomous weapons and battlefield AIs identify and engage threats at machine speed.
- In the courts, virtual reality reconstructions, DNA forensics, and sentencing algorithms analyze evidence and shape judgment.
- Behavioral analytics and lie-detection systems expose deception and flag danger before it erupts.

Protection Miracles

- Biometric security systems guard people and property.
- Body armor, exoskeletons, and autonomous vehicles enhance safety and prevent injury.
- Wearables and rescue drones summon help and save lives in emergencies.
- Air-defense networks and engineered barriers—tsunami walls, surge gates, wildfire breaks—shield populations from large-scale threats.
- Cybersecurity protects digital infrastructure and data integrity.

Prophetic Miracles

- Predictive analytics and machine learning forecast trends in business, health, and behavior.
- Weather, disaster, and disease models predict hurricanes, earthquakes, and epidemics.
- Economic forecasting tools reveal market shifts and guide global strategy.
- Augmented-reality systems anticipate complications in surgery and treatment.

Communication Miracles

- Telecommunications and the internet enable instant global communication.
- Brain-computer interfaces facilitate thought-based communication and brain-to-machine communications—that is, the miracle of telepathy.
- Translation apps and AI models dissolve linguistic barriers between cultures.
- Holographic technology enables in-person, 3D communication.
- VR headsets and haptic feedback teleport users into shared virtual environments.

Victory-in-Battle Miracles

- Cyber warfare cripples enemy infrastructure without traditional combat.
- Directed-energy weapons deliver laser and microwave strikes with pinpoint accuracy.
- Smart munitions and autonomous systems enhance targeting precision and minimize collateral damage.
- AI-driven command networks process data at lightning speed, amplifying tactical advantage.

A Moment of Not-Zen (a.k.a., the Age of Holy Sh*t)

If that catalogue of wonders feels overwhelming, pause and look around. You may not think of yourself as a miracle worker, but by 9:00 most mornings, you've already reenacted half of the Old Testament.

You summoned knowledge from the ether via Google. Moved money with the wave of your hand to buy coffee through Apple Pay. Spoke to a friend across the globe with the touch of a finger thanks to FaceTime. Conjured fire on a smart stove and parted the clouds on a weather app. Maybe you even raised the dead, otherwise known as deepfake avatars.

But that's the issue: We don't call these *miracles*. We call them *Tuesday*.

It's a problem of perspective, and it's nothing new. Futurist Ray Kurzweil used to joke that we don't notice progress in AI because we keep renaming it—like the ATM. Technology has caught up to mythology, but we've misplaced the experience of awe.

Yet as we'll see, awe, and its darker counterpart, holy terror, are essential tools for the journey.

And make no mistake, this is your journey.

You're the one with godlike powers at your fingertips—and you thought getting the kids to school on time was a lot of responsibility.

So consider yourself warned. Because things are about to get weirder.

Exponentially weirder.

The Downside of Up

So, what's exponentially weirder than the miracles that fill our lives? The fact that we barely notice. Humans have superpowers, yet you wouldn't know it from reading the headlines.

Instead, we find dystopia in every direction.

Despite the fact that 78 percent of all companies use AI—and have seen significant gains as a result—the headlines tell a different story. They focus on the threat the technology brings to our economy, our political stability, our very survival. There are similar fears that video games will rot our minds, text messages will destroy our language skills, and robots will steal our jobs. This constant drumbeat of dread has reshaped our collective mood. Mental health disorders are at record highs. According to the World Health Organization, depression alone costs business more than a trillion dollars a year.

Even after five pages of human-created biblical miracles, we don't feel blessed. We feel besieged. Most of us still believe that today is worse than yesterday, and best not think of tomorrow. Spend time watching the Crisis News Network known as CNN, and the message is clear: The end is nigh. When the world changes faster than we can comprehend, we reach for analogies. When our analogies fail, we reach for archetypes. When archetypes fail, we imagine the apocalypse.

When fear becomes our default setting, it colonizes our imagination. Once again, we can see this trend in movies and television. At the same time that archetypal media is on the rise, we've witnessed a parallel explosion in apocalyptic fare. Sure, the genre has always lurked in the background, but the modern wave that began with *The Matrix* and *28 Days Later* has become a deluge: *The Road, Jericho, The 100, World War Z, The Walking Dead*, and all its zombified stepchildren.

The root of this problem runs deeper than failed analogies, exhausted archetypes, or our obsession with the apocalypse. It's not just cultural; it's cortical. The trouble begins in the gray matter inside our skulls. The brain is a prediction engine—literally. In every moment, it's trying to predict what is about to happen and how much energy will be required to meet the challenge. Efficiency is its goal. We evolved in an environment where resources were in short supply. As a result, what the brain wants most is precision prediction, so not a single hard-won calorie goes to waste.

When you approach a door, your brain asks a series of questions: How heavy is the door? Is it open or locked? How much force is required to open it? Is anyone on the other side? When the world matches our predictions, we don't notice. We just walk through the door and keep going. But when our prediction is misaligned with our reality—we think the door is open, but instead find it locked—the brain's alerting network floods stress hormones into our bodies while the voices in our heads shout: "Warning! Door is locked!"

Like AI, the brain uses pattern recognition to make predictions. The goal is to match information from our present to experiences from our past in order to predict the future. But very little in our past has prepared us for the future that is currently unfolding. This is the reason analogies are failing and doomsaying is on the rise.

The environment where humans evolved was local and linear. *Local* means that most everything in our lives was less than a day's walk away. *Linear* means the rate of change crept along, and the differences between generations was often negligible. These are the patterns the brain evolved to match. Today, the world has gone global and exponential. *Global* means that news of events that occur on the other side of the planet arrive in our lives in just moments later. *Exponential* means, forget about the difference between generations, today seismic shifts erupt on a weekly basis. But the brain was not designed to process information at either this speed or scale.

This is why archetypal characters and apocalyptic fictions are on the uptick. These stories help us steady ourselves in the maelstrom. They're the byproduct of the brain's struggle to keep pace with the acceleration of acceleration rendered as media.

At the core of this struggle is information itself—the fuel for both technology and thought. However we define it, information is always seen as more ethereal than corporeal. For the mathematically inclined, it's quantified as bits and calculated as probabilities. For students of information theory, it's the measure of uncertainty in the system. In computer science, it's data—processed, organized, and structured in a way that's meaningful to the user.

Yet none of these definitions explains our struggle.

Information may be ethereal in concept, but it's corporeal in impact. It doesn't just float through the cloud—it rewires the brain. Every bit of data we encounter stirs neurons, excites nerve tissue, and fires electrochemical signals. Action potentials race along axons. Neurotransmitters flood synapses. Genes mutate. Epigenetic cascades unfold. And all this neural noise produces feelings. Emotion is the body's readout of information. There's nothing abstract about it. Information changes us, molecule by molecule and mood by mood.

And this brings us to information's very real corporeal impact: information overload, the wrecking ball of the modern world. Chronic stimulation has exploded into constant stress. It's a blitz our nervous systems were never designed to withstand. The fallout is both predictable and global: depression, exhaustion, burnout—on a planetary scale.

In fact, if you try to quantify the wrecking ball of overload in bits and bytes, the standard metrics of information, well, once again, there are no easy grounds for comparison. In 3000 BCE, for example, if you measured the amount of data in the world by adding up papyrus scrolls, clay tablets, and the like, you totaled out at a gigabyte's worth of data—or the equivalent of four thousand books.

That was then.

If information impacts nervous tissue, let's talk about the impact felt in 2012, the year we published *Abundance*. That year, the world produced 2.8 zettabytes of data. What's a *zettabyte*? It's a trillion gigabytes. That's four thousand trillion books—which is way too many to count and a bombardment the brain was never built to process.

Once again, that was then.

By 2025, the number climbed to 181 zettabytes of information. We don't have an analogy for a byte count that high. And that's our point. Information impacts the nervous system and we're living through the biggest information surge in history. The result is a mismatch between the data storm outside and the prediction engine inside. Our ancient brains don't have the bandwidth and our imagination has been hijacked by the apocalypse. No wonder we don't feel divine.

The Bias Cascade

Information overload doesn't just exhaust the brain and frighten the mind; it invades, occupies, and colonizes attention—our interface with reality and perhaps our scarcest resource of all.

In humans, the bandwidth of attention is 50–120 bits wide. That's all you can focus on at any one time. That's it. Your moment-by-moment reality fits through a channel only a few dozen bits wide.

Listening to someone speak requires about 60 bits' worth of attention. If two people are talking at once while a faucet drips, it's game over. Attention is maxed. You're tapped out.

How tapped out?

As of 2024, the average user spent almost two and a half hours a day on social media. The attention span of a goldfish is estimated at three to nine seconds. The attention span of a human trained on two-plus hours of social media a day is shorter.

So what's the proper analogy?

An insect? No. That's giving us too much credit. A *hungry* insect.

When a bee forages for nectar, it takes one to two seconds to determine if a flower is worth its time. That's how long the modern brain considers an idea before swiping right.

When attention collapses under the weight of too much input, the brain does what it always does—it looks for shortcuts. Our brains try to solve the problem of overload with heuristics: mental rules for processing large amounts of information in short time frames. Common sense is a heuristic. It's a simple rule that turns a heap of data into a straightforward description of the future. Trial and error is another. Yet heuristics don't always work as designed, and they almost never hold up in the face of cognitive overload. When attention is stretched thin, our shortcuts fall apart, and then we call them *cognitive biases*—but really, that's just a fancy way of saying a bad prediction.

Today, even our biases are biased. Our prediction software can no longer keep pace with our accelerated world and our biases layer atop one another, each distorting the next. Exponential bias blinds us to compound growth. Status-quo bias makes us cling to the familiar. Recency bias traps us in the endless now. And together they form a *bias cascade*—a chain reaction that warps reality itself. Consider our negativity bias. In practical terms, this means we prioritize danger over delight. This tilt toward the dour is natural selection doing its job. Humans evolved in an era of immediacy, when threats were of the life-and-death variety. A tendency to see every rustle in the bushes as a saber-toothed tiger was a good way to stay alive. But today, most of the threats we face are probabilistic in nature—the economy might nose-dive, terrorists could attack—yet the brain can't tell the difference. Worse, the amygdala, our danger detector, can't shut off until the threat has vanished. Since probabilistic dangers rarely vanish, our nervous systems remain redlined. *Hypervigilance* is the technical term. Burnout is our lived experience.

The bigger issue is that biases compound. Take our confirmation bias. Faced with information overload, this tendency shifts into overdrive. The brain copes with the deluge by finding facts that align with core beliefs while ignoring all contradictory evidence. This was a culture-building mechanism that helped build stable societies when we lived on the veld. In a world connected by social media that's been algorithmically designed to highlight preferences and bury dislikes? It's a recipe for conspiracy.

Now, layer on our hyped-up negativity bias atop our overtaxed confirmation bias and no wonder we hear so much about disasters and rarely about miracles. Overwhelmed by information, starved for attention, and addicted to apocalypse, we've grown blind to the extraordinary. Yet hidden beneath the noise is a torrent of good news. Miracles are multiplying. Acceleration is accelerating. And the same exponential forces that are unsettling our lives have given us the ability to remake the world—what we call the miracle of abundance.

CHAPTER TWO

An Abundance of Abundance

A Ladder to the Stars

If you want to understand the miracle of abundance, start with the basic function of technology. At its core, technology is a resource-liberating mechanism. It transforms scarcity into abundance through accessibility.

Consider a fruit tree. If you're out picking apples, you only have access to the low-hanging fruit, the stuff you can reach standing on your tippy-toes. If someone hands you a ladder—*Hey, bud, want a ladder?*—the situation changes. Now you can pick the full tree's worth of fruit, and your afternoon snack is plentiful enough to feed your family. Technology—in this case, a ladder—liberated once-scarce resources. It turned a meager harvest into a bountiful haul.

But there are ladders, and then there are ladders. . . .

Take the world's first irrigation systems, another ladder out of hunger. Six thousand years ago, early Mesopotamian farmers faced off against the seasons. When the rains failed, so did their ability to grow food. This wasn't from lack of water. There were ample rivers nearby. Lack of access was the bottleneck.

So when some clever Mesopotamian figured out how to dig

channels from the river to their fields, they didn't just discover a new way to feed their family—they accidentally invented civilization. Suddenly, water flowed year-round. Crops flourished. Surplus food appeared for the first time. Scarcity gave way to stability—and soon after, to society itself. Archeological studies suggest that by 3000 BCE, the populations in regions like Sumer had nearly doubled, fueled largely by the newfound reliability of irrigated agriculture. Canals transformed scarcity into abundance.

Now think about the domestication of the horse. Before horses, moving yourself—or anything else—was slow and effortful. Around 3500 BCE, someone in Kazakhstan figured out how to climb onto the back of an animal strong enough to carry a human at twenty miles an hour. Overnight, trade routes expanded, information traveled faster, and cultures connected. Genetic evidence shows that by 2200 BCE, innovations in equine breeding doubled the number of available animals. This unleashed a mobility revolution across Eurasia. The boundaries of human possibility gained actual, literal, horsepower.

Or the sail. For millennia, the open sea was a barrier. A blue wall that ended the known world. But then another enterprising Mesopotamian, circa 4000–3000 BCE, caught the wind in a piece of cloth. Rivers and oceans became highways. Trade, exploration, migration, culture: All leapt forward. Another ladder. Another liberation. In each case, a simple tool—a trench, a bridle, a sail—took a resource that was once scarce and made it abundant, transforming the impossible into the inevitable.

Today, we are watching this pattern unfold again, only this time at the speed of silicon. You want proof? Meet Thach Ren, a rice farmer in Vietnam's Mekong Delta. Thach has been growing rice in the Mekong for decades, which gave him a front-row seat for a natural disaster. Thanks to climate change, one of the world's most fecund rice locales has begun to dry up, rendering traditional irrigation methods obsolete.

To fight back, Ren partnered with Tra Vinh University and applied technology to the problem. He helped test a smart irrigation system comprised of IoT sensors, smartphone connectivity, and a water management method called *alternate wetting and drying*. Crop yields stayed the same, or rose, while water usage dropped by 20 percent. And that partnership was a pilot study, so those numbers will continue to grow. Our point: Technology turned scarcity into surplus. A ladder, once again.

But we're getting a little ahead of ourselves.

First, try to imagine what those moments must have felt like—the first fields to fill with water, the first horse to haul humans, the first sail to catch wind. The joy of discovery, the thrill of possibility. But more than that, the feeling of safety. Of security. Food, water, transportation, communication—the impossible became the reliable. Imagine what this meant for your chances of survival, for your family's survival. That hope you feel now—that sigh of relief in your soul—this is what we mean when we say technology is a resource-liberating mechanism.

A Decade of Amazing

Every age has its ladders. The stone age gave us shelter, the bronze age gave us power, the steam age gave us motion, and the silicon age gave us intelligence. By the start of the last decade, the reach of those ladders had become exponential. The miracle of abundance was no longer an optimistic idea. It was starting to become a lived experience.

In the decade between 2012, when we explored this trend in our first book, *Abundance*, and 2022, when we started thinking about this one, technology's ability to meet basic needs shifted from debate to data. By the numbers, it was extraordinary. We saw measurable gains in per capita income, access to food, energy, communications, education, and healthcare (we lay out the Proof of Abundance in appendix A). These are core metrics of prosperity. These jumps caught our attention. Sure, we were the ones who made those rosy predictions back in 2012. But it was a decade of amazing progress, and watching the exponentials at work was something to behold.

In that span, over two hundred million people escaped extreme poverty—despite COVID-19 setbacks. More than a billion people gained access to electricity. Over two billion gained access to safe drinking water.

And those historic leaps were dwarfed by the spread of communications abundance. In 2012, 2.4 billion people were online. By 2025, the number exceeded 5.5 billion. In less than a generation, 3 billion people joined the global conversation.

But computing takes the cake. In 2012, a billion people owned a smartphone. By 2025, the number was over seven billion—close to 90 percent of the planet. In *Abundance*, we calculated that the average 2010 smartphone contained over a million dollars' worth of 1980s technology: phone, camera, computer, GPS, encyclopedia, music players, music library, and more. Using this same benchmark to gauge progress between 2010 and today—including processing power, camera quality, connectivity, and new features like health monitoring, augmented reality, and AI assistants—we see an exponential increase in "value density" worth $7.1 million. If we use access to capability and not income as a metric for wealth, then nearly seven billion people became multimillionaires over the past decade.

That's roughly $49 quadrillion in newly created global wealth.

And that's today. Tomorrow, consider the two technologies we examined in the previous chapter: AI and robotics. Today, warehouse robots and autonomous cars are remaking manufacturing and transportation. Tomorrow, these same capabilities will democratize healthcare and education, serving rich and poor alike. Labor markets will be transformed by robot workers who operate 24/7—no sick days, no burnout, no need for pay raises.

As venture capitalist Vinod Khosla pointed out in the newsletter *Thoughts from the Frontline*: "In twenty-five years, there could be a billion bipedal robots. . . . [This] could free humans from the slavery of the bottom fifty percent of really undesirable jobs. This will radically change GDP, productivity, and human happiness. These robots could create enough value to support the people they replace."

The more important point is that abundance compounds. Access to clean water doesn't just slake thirst, it reduces waterborne diseases, which improves child survival rates, liberates women from backbreaking labor, and spurs entrepreneurship and economic growth. The smartphone keeps us in touch and also unlocks mobile banking, microloans, and online education. Together, these forces create a flywheel of progress, spinning ever faster.

Yet there are costs. An abundance of screen time has led to an abundance of mental health challenges: An abundance of ultra-processed food has led to an abundance of obesity and metabolic illness. An abundance of social media has led to an abundance of polarization in society—the tribalism of online echo chambers amplified by the acceleration of fake news. And, of course, an abundance of artificial intelligence has spurred an abundance of fears: hostile takeover by a rogue machine, massive job losses produced by AI-powered robots, bad actors with malevolent intentions and supercomputers at their disposal.

Abundance is a double-edged sword. There's a vast upside and a very real downside. Our conclusion is simple: Abundance is our future, but it won't be the future humans want unless we learn the lesson of Spider-Man, that with great power comes great responsibility.

Unless we can learn to mitigate the unintended consequences of abundance, the benefits of accelerating technology may not be enough to safeguard civilization. And the clock is ticking. So, how to create the future we want before it's too late?

This is the focus of *We Are as Gods*. It's a blueprint for thriving in an accelerated era: a deep examination of where we are now, where we're going, and what to pack for the trip.

Our argument is not that abundance is coming; it's already here. Our argument is that abundance is scaling, in ways that boggle the mind and threaten our capacity to cope. The question is how to harness this bounty without being crushed by its weight.

While our outlook is optimistic, we do not shy away from the troubling implications of accelerating technology. Economic issues such as job loss and worker reskilling. Emotional issues such as anxiety and depression. Existential issues from disaffected youth to misaligned AI. What can be done to mitigate these concerns?

To answer this, we provide a series of psychological interventions based on neurobiological mechanisms designed to reprogram the brain for speed and scale, the hallmarks of our era. Our mission is a manifesto for a better tomorrow. It's a call to action. How to seize the extraordinary opportunities before us, navigate the very real challenges we now face, and not lose our minds in the process.

The lessons that follow are grounded in science, sharpened by experience, and built for action. They prepare you for the adventure that is a world of everything, everywhere, and all the time.

Where to start?

In an age of information overload, our most vital tool is discernment.

Ruthless discernment.

Truth Filters

Until recently, we outsourced discernment. Journalism, the field where Steven started his career, is a good example. In the 1990s, reporting for major newspapers and magazines, Steven was trained to treat facts as earned, not assumed. The core question: How do you know a fact is a fact?

The answer: the standards of journalism. A reporter could print a claim only if three independent sources corroborated it. If a scientific result appeared in a major journal—*Nature* or *Science* carried more weight than, say, *Frontiers in Psychology*—and three unaffiliated experts could verify it, then it made the page.

Next came the gauntlet. A fact-checker called every source. An editor combed the copy, then a managing editor, then the editor in chief. In the old model, any fact in a credible news outlet had been cross-examined by a half-dozen people professionally trained in ruthless discernment.

Those days are over.

Communications abundance eroded this rigor. Anyone with Wi-Fi and a keyboard can "break" the news. Zero requirements for the job. No apprenticeship. No gauntlet of editors. No fact-checkers guarding the castle. The gatekeepers are gone.

Yet there's no free lunch. With great communications power comes great communications responsibility. In an age of abundance, the responsibility for fact-checking has shifted from producer to consumer. What we need now are *truth filters*—ways to reliably assess the validity of information.

The standards of journalism are a truth filter. The three-sources-per-fact rule provides a metric for accuracy. It doesn't guarantee capital *T* truth—very little does—but it yields an educated consensus and an audit trail for how that consensus was reached.

Science is another truth filter. The scientific method tests hypotheses against evidence, discards errors, and converges on repeatable results. Falsifiability draws a clear line: If a claim can't be disproved in principle, it isn't scientific. Occam's razor favors simpler explanations when evidence is equal. Probabilistic reasoning updates belief with new data. The Socratic method stress-tests assumptions. Each is a battle-tested protocol for unearthing truth.

And never have we needed them more.

Today, you can't outsource accuracy. Each of us must become our own gatekeeper. The castle we guard is our mind.

Truth filters are essential for thriving in an age of abundance—and for the argument that follows. To ground our case, we'll run the evidence for abundance through a few of the most powerful truth filters yet invented. The first is *first principles thinking*, the foundation of scientific reasoning. The second is the *Six Ds of Exponentials*, a framework that tracks how exponential technologies evolve from deceptive to disruptive. By combining these truth filters with insights gleaned from entrepreneurs using these same technologies to tackle global challenges, we'll test our central claim: the evidence for abundance at scale.

The First Principles of Abundance

A long time ago, back when the Greeks wore sheets, a sheet-wearing Greek named Aristotle penned a book called *Metaphysics* and introduced one of the most famous truth filters in history: first principles thinking. This is defined as the process of identifying self-evident truths—those that cannot be deduced from anything else. As Aristotle wrote: "the first basis from which a thing is known."

Like all truth filters, first principles thinking has been stress-tested by two of the most brutal forces in history: time and genius. Across the centuries, this same logic shows up everywhere. During the Renaissance, by advocating for systematic observation and reasoning, Francis Bacon gave us the first principles of the *empirical method*, while René Descartes claimed "*Cogito ergo sum*" (I think, therefore I am) as the first principle of knowledge. Next came the Industrial Revolution, where James Watt reduced mechanical systems to fundamental physics and Michael Faraday laid the groundwork for electric motors from the first principles of electromagnetism. In the twentieth century, Albert Einstein leaned on first principles to challenge Newtonian assumptions. Henry Ford broke car production into elemental tasks. Richard Feynman redrew quantum mechanics, where his diagrams turned first principles into visual proofs. Then Steve Jobs imported the idea into Apple, to simplify design and product development.

Most recently, Elon Musk's career has become a case study in first principles. To build reusable rockets, Musk didn't inherit expensive designs from the aerospace giants. Instead, he asked: *Why are rockets so expensive?*

Answer: *Because they are only used once, then thrown away.*

Next question: *What would it take to make them reusable?*

Answer: *Multiuse engines and a rocket able to survive space travel, with powered reentry and landing capabilities.*

So Musk got to work. He reengineered the Merlin engine and redesigned the first stage of the Falcon 9 rocket to include landing legs. The results slashed costs tenfold.

Next, he applied first principles to electric vehicles. Dismissed as impractical and expensive by the auto industry, Musk asked: *What stands in the way of widespread adoption?*

Batteries.

At the time, batteries were heavy, dangerous, and expensive.

But why?

Next question:

What's in a battery?

The answer: *Lithium, cobalt, nickel, graphite, aluminum, and copper.*

So Musk checked prices on a metals exchange. Each of these materials cost pennies on the dollar. All the expense lay in production. By reimagining battery design, manufacturing, and supply chains, Musk disrupted the automotive industry and accelerated the global shift toward sustainable energy.

But first principles thinking isn't just a tool for entrepreneurs—it's a way of seeing. It cuts through assumptions and clears away our biases, helping us examine the evidence for abundance at scale.

Start with energy. Energy is not a scarce resource, it's just hard to harvest. Every hour, the earth receives more solar power than humanity uses in a year. The sun provides energy abundance; our challenge is how to harness, store, and distribute it. This is why Musk focused on batteries: Sunlight is free. That's the first, first principle of energy abundance.

Water scarcity follows a similar pattern. Seventy-one percent of the earth's surface is covered by oceans. In fact, scientists recently discovered a mega-ocean buried four hundred miles beneath North America the size of all the planet's surface oceans combined. Once again, resources are not the problem.

The problem is that 97.5 percent of our surface water is salt water, while that mega-ocean is bound in mineral crystals and not easily accessible. The bottleneck is harvesting, purification, and distribution.

Yet solar-power desalination costs have dropped 60 percent in the past decade. In the Middle East and North Africa, desalinated water is cheaper than traditional sources. The XPRIZE (Peter's organization), in partnership with the leadership of Abu Dhabi, just launched a $119 million competition to slash desalination costs even further, while making them more sustainable.

And who needs desalination when there's air?

The atmosphere contains thirteen sextillion liters of water. More than enough to meet our needs. Atmospheric water capture allows us to harvest that bounty, even in low-humidity environments. In 2024, Florida deployed these devices to supply hospitals during Hurricane Milton, while large-scale projects, like the $26 million Hawaiian initiative to equip a thousand homes, are already underway.

Let's take this thinking one level higher and consider the intelligence that powers these innovations. From a first principles perspective, intelligence is our most critical resource. Welcome to 2025, where intelligence—in the form of generative AI—just went exponential.

Everything around us—bed, desk, chair, car, toilet—is getting smarter. We are embedding intelligence into robots. By 2040, ten billion humanoid robots are expected. Their anticipated cost: $20,000–$30,000 per unit.

If these numbers are correct, you will be able to lease a robot for about $300 per month, or $10 per day, or $0.40 per hour . . . or about fifty times cheaper than today's minimum wage.

What does this mean?

Do you shovel snow in the winter? Not for long.

Do you hate doing laundry? Not for long.

That's the point.

A decade back, exponentials had invaded our routines but not yet transformed our lives. A decade later, we're at scale. And speed. Today, abundance is visible everywhere we look, whenever we look. If we apply first principles thinking to how we look at the next ten years, we see . . .

To show you what we see, we must first introduce another truth filter: the Six Ds of Exponentials.

The Return of the Six Ds

First principles thinking breaks the universe into fundamental truths. It's a tool for invention and insight. But understanding technological progress requires more than a microscope; it requires a *macro-scope*, a time-lapsed view of emerging breakthroughs and compounding abundance.

Enter the Six Ds of Exponentials: a truth filter that reveals how disruption unfolds and transformation scales. While first principles thinking identifies the building blocks, the Six Ds trace the chain reaction that turns innovations into revolutions. From digitization and deception to dematerialization and democratization, it's a blueprint for technological development and our means of predicting the future. And once you see this cycle, you start seeing it everywhere.

We introduced the Six Ds in *Abundance*. Now, a decade later, we return to them—updated, expanded, and more urgent than ever.

Forget Moore's Law. This is about the Law of More.

Stage One: Digitization

Anything that can be turned into the ones and zeroes of computer code can be digitized—that's the first *D* in the cycle.

Once that happens, hold on. Digital information moves at the speed of light: It can be copied, shared, and improved endlessly for almost zero cost. Once something can be represented by code—from math and music to biotechnology and blockchain—it becomes an information-based technology that jumps onto the back of Moore's law and starts to accelerate.

In 2012, digitization only reached the low-hanging fruit: industries already built on data. Entertainment and communication led the charge. Music, movies, books, and photography went digital nearly overnight, while E-commerce, online banking, and early cloud services gained ground. Yet AI stumbled over basic tasks like speech and image recognition, physical processes such as manufacturing and healthcare remained resistant to this trend, and the more complex domains of emotional experience and biological systems? Still off the map.

A decade later, digitization reaches everywhere. AI can interpret the world. Appliances, factories, and whole cities now speak through the

Internet of Things. Digital twins mirror everything from jet engines to the human body. Wearables and genomics translate biology into data, while emotion-recognition systems read moods and mental health. Even creativity has crossed over. Generative AI now writes, paints, and composes. What once belonged to the realm of bits now saturates the world of atoms. And once this crossover happens, everything built on top of bits starts compounding. Once information becomes code, that code begins to eat the world. And then—nothing. Exponential growth disappears from sight. That's the next stage of the Six Ds cycle: deception—when the miracle of progress hides in plain sight.

Stage Two: Deception

Progress hides in plain sight because of simple math. Our linear brains can't track the doubling of small numbers. Start with 0.01, double it, and get 0.02, then 0.04, then 0.08—and who beyond experts would notice these changes. This is the deceptive phase of exponential growth, the second stage in the cycle, and the calm before the explosion. Once those numbers cross the whole-number barrier, they rocket into the billions in an eyeblink. Start with a penny on January 1 and double it daily—one cent becomes two becomes four—and you're a billionaire before February.

For a real-world example, consider COVID-19. Patient zero appeared in Wuhan, China, on December 1, 2019. A month later, the World Health Organization reported forty-one cases. By January 20, 2020, there were 282, with the first international cases appearing in Southeast Asia. At that point, the virus was doubling every two and a half days. Exponential growth was already underway—just invisible to most. By March 2020, the virus had reached sixty-one countries and eighty thousand people. One month later, one million. By May, five million. By June, ten million. Fifty million by November. A hundred million by January 2021, one year from the start. The imperceptible had become the unstoppable. That's the signature of deception.

In business, the same curve kills giants. Kodak's engineers invented the digital camera in 1975, but executives dismissed it as a toy. They missed the exponential doubling of pixels, storage, and computing power that would soon make film obsolete. By the time the trend broke the surface—when every phone became a camera—Kodak was bankrupt.

Hollywood made the same mistake on a global scale. In the mid-twentieth century, it controlled 70 percent of the world's content market. Then came YouTube—an exponential force that blew open the gates. No studios, no agents, no barriers to entry. Anyone could create, upload, and share, for free.

During its peak decade, Hollywood produced about 1,150 hours of new content each year. YouTube now adds 500 hours every minute—a quarter-billion hours annually, roughly 230,000 times Hollywood's output. In 2022, YouTube earned $29 billion in ad revenue. In 2026, analysts project it will surpass Disney's $54 billion, making it the biggest media company in the world. Did anyone see this coming? That's the point. The deceptive phase is deceptive. It blinds companies and sucker-punches industries. But once that exponential curve breaks into sight, buckle up. Disruption has arrived.

Stage Three: Disruption

Disruption is when exponentials kick down the door. The invisible becomes visible as new products replace old products and traditional markets vanish while novel ones emerge. If you stream music, why buy CDs? If you snap photos with your phone, why buy a film camera? If you have ChatGPT, why bother with traditional search engines?

When we wrote *Abundance*, disruption meant a shift from analog to digital—streaming movies via Netflix replaced renting DVDs from Blockbuster. Today, it cuts deeper. Technologies like 3D printing, AI, and advanced robotics are invading industries once considered untouchable. In healthcare, wearables track the body in real time, while generative AI accelerates drug discovery. In logistics, robots, autonomous trucks, IoT networks have reinvented how goods move. And with 3D printing enabling local production, outsourcing—and the nightmare of tangled supply chains—is fading fast. Yet the biggest difference of the past decade is disruption's ability to meet our basic needs. In ten years' time, exponentials disrupted the core systems of survival: food, water, energy, healthcare, and education. The Proof of Abundance appendix tells the full story, but here's the short version: AI and robotics transformed agriculture through precision farming and lab-grown meat, and millions no longer go hungry as a result. 3D printing and modular construction cut build time and housing costs, so shelter needs can be met faster and cheaper than ever before. Solar and wind replaced coal and oil, making renewable energy accessible and affordable. How disruptive is disruption? In a decade, abundance for all has become business as usual.

And once disruption occurs, we move to demonetization—when cost itself gets removed from the equation.

Stage Four: Demonetization

In 1991, if you wanted to send a message across the world, you needed an international calling plan, a fax machine, or a stack of stamps. Ringing your cousin in Cairo was a special occasion. Long-distance romance was a money pit. Communication used to cost.

Today, a five-year-old with Wi-Fi can video chat with Grandma on the other side of the planet for free. What once required a fortune is now a feature. That's the power of demonetization. It's the stage in the cycle when cost fades into the background. Here, the cheap becomes cheaper, often free.

Take computing. In 1976, the Cray-1 supercomputer was the world's most powerful machine. And it was a machine: weighing 10,500 pounds, drawing 115 kilowatts of power, and costing $5 million—about $25 million in 2025 dollars. Compare that to the iPhone 16, which weighs 16 grams, uses 1–2 watts of power, and costs $799. In the past fifty years, the forces of demonetization shaved $24 million plus off the price tag of supercomputing.

This same force is now reshaping our ability to meet basic needs—energy, education, and healthcare. Since 2010, solar saw an 85 percent price drop. It's one of the cheapest energy sources available today. Online platforms like Khan Academy and Coursera have demonetized learning, offering no-cost or low-cost access to top-tier instruction. Telemedicine and AI-powered diagnostics now bring affordable healthcare to millions.

We're saving money and unlocking potential. Demonetization creates a positive feedback loop for thriving. More people with access to energy, education, and healthcare means more minds have joined the global conversation—and more innovation follows. The feedback feeds back. As innovation spreads, economies grow, and standards of living improve.

Stage Five: Dematerialization

Now you see it, now you don't—welcome to dematerialization, the next stage in the cycle. In the previous stage, cost vanished from the equation. In this one, physicality itself disappears. Once you bought stereos and GPS systems and video cameras and all the rest. Now you buy a phone, and all the rest comes standard.

But check out the gritty details.

Compare the 0.37 pounds iPhone 16 to the 1990s technology it replaces. What was once a Canon EOS 1 SLR (2.9 pounds), a VHS camcorder (5 pounds), a Motorola bag phone (4.5 pounds), a Macintosh Classic (16 pounds), the *Encyclopedia Britannica* (129 pounds), a Discman and boom box combo (10.6 pounds), a Garmin GPS (1.5 pounds), a portable CRT TV (10 pounds), and, of course, a Nintendo Entertainment System (6 pounds)—now sits in the palm of your hand.

We see the same disappearing act in the rise of "as a Service," or aaS. The trend began in the early 2000s, when Salesforce launched Software as a Service. Soon after, Amazon and Microsoft turned cloud computing into Infrastructure as a Service. The 2010s added Platforms and Hardware as a Service, followed by AI and Blockchain as a Service in the 2020s. Today, it's all collapsed into a single acronym: XaaS—Everything as a Service. The pantheon now spans Mobility as a Service (autonomous taxis) to Biology as a Service—better known as Biotechnology as a Service—where CRISPR gene-editing and DNA synthesis are outsourced to the cloud. Taken together, these shifts mark the business-model side of dematerialization.

Today, dematerialization is transforming our ability to meet basic needs. Stadium-size coal-fired power stations are being replaced by solar panels on our roofs. Libraries disguised as e-readers fit inside our pockets. Room-size diagnostic tools now live in smartwatches. Dematerialization makes the digital intangible. Less raw material is consumed on the front end. Less environmental strain on the back end. And in the middle, a radical increase in accessibility.

Stage Six: Democratization

Riddle me this: It's 2010. You're on a game show, sort of like *The Price Is Right*, only for future technology. Here's the puzzle: A decade from now, there will be a technology that can make movies; write scientific papers; code apps; draw up contracts; balance your checkbook; edit your bestseller; diagnose illness; and, by the way, tell you how to cook dinner when all you have in the fridge is a jar of mayo, some peanut butter, and an old rice cake.

Who would have access to this technology? The rich and powerful? Silicon Valley billionaires? Heads of state?

And for what cost?

A million dollars a month? Maybe more?

But as digitization and deception give way to disruption, demonetization and dematerialization culminate in democratization. In this stage, technology goes wide. ChatGPT, the above example, is the most powerful technology in history. It's now available to all of us for free—though not free of carbon.

It's not the only example. Smartphones were once luxuries for the rich. Now, they serve everyone, everywhere, regardless of class. Internet access followed suit, with broadband being beamed to even the remotest corners on the planet. Ride-sharing apps have replaced car ownership. Even expertise is being democratized. AI allows anyone with a Wi-Fi connection to solve problems once reserved for PhDs, give or take the occasional hallucination.

We see the same shift in the technologies of abundance. Agriculture has been transformed by precision farming, 3D printing decentralized manufacturing, and modular solar brought energy independence to many. Even healthcare, once a fortress of exclusivity, has been democratized—wearables offer diagnostics, AI chatbots provide mental health support, and telemedicine connects patients to doctors.

So where does this lead?

If you apply the framework of the Six Ds to the evidence for abundance, we see an exponential promise fulfilled in the world of today. But frameworks only get us so far; to understand what abundance actually feels like, we have to zoom in—from the high-level view of mega-trends to the boots-on-the-ground experience of abundance at scale—a.k.a., daily life in the 2020s.

CHAPTER THREE

Data-Driven Optimism

The Great Blood Bank in the Sky

In 2016, a hospital in rural Rwanda sent an emergency text. A woman was hemorrhaging during childbirth. *Send help. Send type O negative. Hurry.* Only, not in so many words.

Minutes later, a drone was launched. Ten minutes after that, thirty feet above the hospital, it released a small package on a parachute. The package dropped to the landing zone and the doctor received a text: Type O negative was in the house. The new mother got to hold her child. They both got to live. And everything in this story is both a healthcare miracle and a logistics revolution—one that cut maternal mortality in Rwanda by 51 percent.

Miracles of this sort are the focus of this chapter. In the previous one, we went macroscopic. In this one, we shift from the many to the few to examine abundance at scale, meeting entrepreneurs wielding exponential technologies on the front lines of humanity's basic needs. Our focus is a string of real-world interventions: the on-the-ground impact these innovators are having in the communities they serve.

That woman and her child are alive because of Zipline, the largest drone-delivery service on the planet. Founded and run by Keller

Rinaudo Cliffton, Zipline is also a great demonstration of the scale of impact a single entrepreneur can now have. Since launching in 2016, the company has brought a healthcare revolution to Rwanda, has its sights on Africa, and soon the rest of the world. And more impressive than the scale of Clifton's impact is the arena it's unfolding: medical supply delivery—one of the most mind-bending logistical puzzles in healthcare.

But we're getting a little ahead of ourselves.

This story doesn't start in logistics. It starts in robotics, where Cliffton's career began. In college and graduate school at Harvard, Cliffton's focus was on building nanoscale robots that could deliver lifesaving medicines to patients. Afterward, he went in a different direction and founded a company called Romotive, which—in another example of how exponentials drive down price and boost performance—produced an iPhone-controlled robot. "[This] is about one percent of what these robots have cost in the past," Cliffton explained in a TED Talk—as a $150 Wi-Fi-capable, computer vision–enabled bot wandered around the stage.

In 2014, Cliffton found his real mission. He pivoted his company toward a different form of robotics: drones and, more specifically, the use of electric autonomous drones to deliver medical supplies on demand. Getting treatments to hospitals has always been perilous: lack of roads, refrigeration, and electricity contributes to 5.8 million preventable deaths a year. In large parts of Africa and Asia, 70 percent of the roads are impassable in the rainy season. In rural areas, one in three deliveries are delayed or fail outright. If drones could solve this problem . . . but it was a big if.

When Zipline was founded in 2014, most commercially available drones could barely lift a bag of sugar, had a range of less than a football field, and were capable of twenty minutes of flight time before the batteries died. But drones sit at the convergence of robotics, batteries, artificial intelligence, and material science. Cliffton knew exponential upgrades were coming. Plus, the mission kept him going. "When we were [just launching the company]," Cliffton explained on

Peter's *Moonshots* podcast, "someone asked me, 'What are the chances that we'll succeed in building this?' I said, 'Well, they're less than five percent.' People were . . . shocked, disappointed, and started asking, 'Why are we here?' I said, 'But it's five percent of getting to build crucial infrastructure for humanity. That's such an important thing. It's totally worth it.'"

Zipline launched in 2016, with one hospital and one product: blood, which is medicine's hardest delivery problem. Blood has a short shelf life, complicated storage requirements, comes in a huge variety of types, and with a wildly unpredictable demand. Pre-Zipline, Rwanda, a country of fourteen million people, was served by four regional blood banks. Delivery could take days, so doctors often had to gamble: Do they use their limited blood supply for the patient in front of them or save it for another emergency? Zipline's promise was profound: to take the guesswork out of saving lives.

Yet, in the beginning, nobody would invest—except the government of Rwanda. They became Zipline's first customer. And Zipline delivered. And delivered.

Today, Zipline supplies 75 percent of Rwanda's blood. Orders are placed via text, flown via drone, then parachuted to the ground. The package descends to the same landing spot every time. Doctors get alerted upon arrival. Instead of hours or days or not at all, Zipline's blood arrives, on average, twenty to thirty minutes after the order is placed.

It's not just Rwanda reaping the healthcare benefits.

Zipline is now the planet's largest autonomous delivery network, valued at over $4 billion. The company serves over four thousand hospitals in nine countries, including the United States, Australia, and Japan. In total, their drones have logged over 80 million miles and made over 1.1 million deliveries. In Ghana, they took over the delivery of vaccines—over sixteen million doses so far—reducing missed vaccinations by 42 percent and waste by 60 percent. A 2022 *Nature* study found that small drones cut emissions per parcel by up to 84 percent and consume 94 percent less energy per delivery. But if you're looking for a statistic that captures the scale of impact a single entrepreneur can have, consider our first practical miracle: that 51-percent reduction of maternal mortality in Rwanda.

That's data-driven optimism. The next time disaster headlines make you think the world's getting worse, anchor here—because a handful of engineers just halved maternal deaths ten thousand miles from Silicon Valley. If that doesn't redefine the limits of impossible, what does?

The Slaughter of Slaughter

If you're tallying the evidence for abundance by measuring the impact a single entrepreneur can have, Keller Cliffton is a tough act to follow. But what's more impactful than preventing maternal mortality at scale? Preventing ten million deaths a day.

Ten million—that's how many animals are slaughtered every day to feed our planet. That's not a typo. Some ninety billion *land* animals die each year so our species can eat lunch. Add in fish, those figures nearly double. It's an ethical and environmental disaster, and it's getting worse. If the planet's population climbs above the nine billion projected by mid-century, meat demand will increase by 70 percent (compared to 2005 levels). We don't have the resources to solve that problem using traditional methods—nor should we have the stomach for it.

Already, one-third of the planet's arable land is filled with the corn and soy fields required to feed the animals we eat. Livestock uses 80 percent of the planet's agricultural land yet produces only 18 percent of its calories. Water consumption rates are more alarming. It takes around twelve thousand liters of water to grow a kilogram of meat. There's enough water in an adult steer to float a US Navy destroyer. Add in greenhouse gas emissions, where meat production dwarfs all forms of transportation combined, and the industry becomes unsustainable.

What's more, considering all the devastation, you would hope the industry was doing its job—that is, feeding people. Yet, while the majority of abundance metrics are trending in the right direction, this is not true for food. Between 2012 and 2022, the number of people living with moderate-to-severe food insecurity went from under a million to over two billion. These numbers tell us that incremental change isn't going to solve this problem. We need complete overhaul of how we feed the planet.

This brings us to goldfish.

Josh Tetrick was in law school when he heard about the goldfish. Born in Alabama in 1980, Tetrick graduated from Cornell with a degree in Africana studies, then a Fulbright fellowship took him to Africa, before he landed at the University of Michigan Law School. He heard about the goldfish in constitutional law—just this weird NASA paper his professor mentioned in passing.

The paper was about the longstanding problem of feeding astronauts in space. Freeze-dried fare for months on end isn't great for health or mood. In 2001, in an attempt to solve this problem, German scientists tried to engineer an artificial ecosystem that could supply astronauts with fresh fish. Then NASA decided to see if they could accomplish the same feat, minus the ecosystem.

NASA scientists sliced a large chunk of muscle off a goldfish, put it inside a bioreactor, and fed it fetal bovine serum, the same nutrient-solution used to cultivate cell lines. In a week, their filet grew by 14 percent. It was the first glimpse at the possibility that would later be called *cultured meat*.

To Tetrick, the idea of growing meat in a lab seemed like science fiction. Yet he was already concerned about food security. As a Fulbright fellow in Nigeria, Tetrick noticed that people subsisted on vegetables, except on payday, when they splurged on meat. He was in Nigeria to promote education and end the cycle of poverty. If successful, as people rose out of poverty, their desire for meat would rise with them. In other words, Tetrick noticed an "abundance paradox"—that is, the more we succeed at solving one problem (poverty), the greater strain the solution puts on another (food security).

But were goldfish filets the solution? Unlikely. The real opportunity was hiding in plain sight. Tetrick parked the idea of cultured meat and—so much for that law degree—turned his attention to eggs. Like meat production, eggs have staggering inefficiencies. Chickens produce vast quantities of waste, emit a bevy of greenhouse gases, and require enormous amounts of feed, water, and energy. So Tetrick took a page out of NASA's playbook, and decided to see if he could remove the chicken from the equation.

He founded Eat Just and developed a plant-based egg substitute made from mung beans that requires 98 percent less water and 86 percent less land than eggs. Since launch, they've replaced five hundred million chicken eggs that would have taken over a million chickens to produce. Josh Tetrick prevented a slaughter—and he was just getting started.

The goldfish never left his mind. He realized that all the components of cultured meat were available in other industries. Culturing cells from antibodies was standard in vaccine production. Biopharma had nailed cell line development. "And food production," Tetrick explained in an interview on Peter's *Moonshot* podcast, "is really good at taking a raw material and converting it to a finished product."

In 2017, Eat Just blended these processes to create one of the first lab-grown chicken products to receive regulatory approval. It debuted in restaurants in the United States and Singapore in 2023. But Tetrick wasn't satisfied. Beef is still the goal.

Currently, Eat Just is developing CRISPR-enabled cell lines—another process borrowed from biopharma—to enhance meat production. Scaling up bioreactors to grow beef at scale is the next issue, yet exponentials remain in play and market forces are at work. Consider that the first lab-grown burger cost $330,000 in 2013. Today, it's down to ten dollars a kilogram in pilot production. Can Tetrick bring it to scale?

Somebody will, and probably sooner than expected. The opportunity is enormous. Meat is a trillion-dollar industry and the abundance paradox isn't going anywhere. Between the dollars to be earned, the death to be avoided, and the chance to feed our growing world, the writing is on the wall—and the Six Ds put it there.

The Diagnostic Divide

Mary Lou Jepsen grew up on a small farm in Connecticut. Her father was an auto mechanic who taught her how to fix things—and if you can fix things, you can build things. And build things is exactly what Mary Lou Jepsen did.

She studied art and electrical engineering at Brown, went to graduate school in holography at the MIT Media Lab, then returned to Brown for a PhD in optical physics. Along the way, she played in a punk rock band, met icons like Andy Warhol, Lou Reed, and David Byrne, and started her first company. She also created the first holographic video display, built the largest ambient holographic display ever as installation art, and came up with a plan to project images onto the moon (the tech worked but was never deployed).

And then Jepsen got sick.

Actually, she was already sick.

Since the age of twelve, Jepsen had been in and out of the hospital with mysterious illnesses. In her late twenties, she was sleeping twenty hours a day and spending her waking hours confined to a wheelchair. Half her face was paralyzed, so she drooled. Nobody could figure out what was wrong. The worst part was that Jepsen lost the ability to do subtraction in her head. This is when she decided she didn't deserve a PhD in physics, dropped out of graduate school, and went home to die.

In a profound act of kindness, one of her professors sprang for an MRI—which she could not afford. The scan revealed a brain tumor. Her family was horrified. Jepsen was thrilled. She'd been sick for over a decade. That scan saved her life and explained the mystery. Plus, surgery was a solution.

After surgery, Jepsen returned to grad school, finished her PhD, spent a few years as the chief technology officer for Intel's display division, then went back to MIT to become a professor at the Media Lab. There, she teamed up with Nicholas Negroponte to co-found One Laptop per Child (OLPC) and build a low-cost, low-power laptop for children in the developing world. At the time, laptops cost about $1,000. By combining converging exponentials, OLPC drove that down to $180—helping launch the tablet revolution and transforming the lives of millions.

Yet Jepsen kept thinking about the fMRI that saved her life, the one she couldn't afford. "My experience with One Laptop per Child," she explained in an interview with the authors, "showed me how accessible technology can level the playing field. When we remove cost barriers and democratize access, entire communities benefit. Seeing what technology could do for education ignited my passion to apply those same principles to healthcare."

At OLPC, a portable tablet bridged the digital divide. Jepsen's question became: Could a different technology bridge the "diagnostic divide," the difference in access to medical diagnostics between wealthy and impoverished populations? In the United States, for example, there are about 13,000 MRI machines. In Mexico, by comparison, there are about 440.

And it's not just Mexico.

According to a 2021 *Lancet* study, 47 percent of the world's population lacks access to basic medical diagnostics. It's 75 percent, if you include high-resolution imaging. Narrowing the diagnostic divide for only six conditions—diabetes, hypertension, HIV, tuberculosis, plus hepatitis B and syphilis for pregnant women—would reduce premature death in lower- to middle-income countries by 1.1 million lives a year.

Jepsen's question has a Six D answer. In recent years, medicine has gone digital, weathered deception, and witnessed disruption. Portable diagnostics like handheld ultrasound machines, rapid blood tests, and mobile lab kits have spread like wildfire. Wearables now monitor everything from heart rate and blood pressure to sleep cycles and stress biomarkers. Telemedicine demonetized and democratized expert diagnosis. Today, AI is accelerating democratization across the board—already detecting everything from cancers to cardiovascular diseases to rare genetic disorders with remarkable precision.

Yet, despite these breakthroughs, MRI remained stubbornly resistant to change. Around 2016, Jepsen decided to see if she could solve that problem. She founded Openwater in an attempt to make high resolution imaging affordable, portable, and scalable. She wanted a 1000x improvement in the technology.

It was quite a goal.

To create OLPC, Jepsen had to combine three or four exponential trends into a single laptop. What she didn't have to do was dematerialize an entire room's worth of equipment, including the helium-cooled, two-ton superconducting magnet that sits at the heart of an MRI machine. But that's exactly what she did.

Openwater uses near-infrared light and miniaturized ultrasound to shrink the imaging capabilities of an MRI machine to the size of a credit card, at a fraction of the cost. Infrared light penetrates the skin and is cheap to produce—your smartphone's flashlight does the job. The issue for diagnostics has always been scattering. But send an ultrasonic ping through the tissue, and you can focus the beam. The combo allows Jepsen to blend MRI, ultrasound, and near-infrared imaging to create a device that sits inside a headband.

Why a headband?

Because Jepsen's target is brain diseases.

First up: strokes, the number two killer in the world. Strokes are a plumbing problem. Already, doctors know how to dissolve the clots and repair the internal bleeding they produce. The issue is time. Every minute after a stroke begins, nearly two million neurons die. Yet patients present with a dozen symptoms. The subject is slurring their words—are they drunk, or did they have a stroke?

Openwater's headband solves this by providing portable brain imaging at the point of care, identifying clots or bleeding in seconds, and drastically reducing the time to treatment. If you can get a stroke patient from diagnosis into surgery in under two hours, not only do they live, they have a 90 percent chance of no neural deficits.

Strokes are the starting point. Next, Openwater's going after brain cancers. Their first target is glioblastoma. The disease is almost always fatal because the cancer cells hide in tangles of neurons, so enmeshed in the tissue that removal is impossible. Yet cancer cells are fragile. Low levels of focused ultrasound can burst their walls and leave surrounding tissue unaffected. And proteins from the exploded cell vaccinate the brain against that particular cancer. In preclinical trials, the technique performed five times better than chemotherapy.

After brain cancer, it's mental health. Turns out, the same frequencies that shake apart cancer cells can also stimulate neurons. This can trigger the birth of new neurons, which could improve memory, or govern the release of neurotransmitters, which could improve mood. This makes Openwater's device a potential treatment for Alzheimer's, PTSD, OCD, addiction, depression, and anxiety, and one of the most disruptive advances in the history of medicine.

Another reason for that disruption? Openwater is going open source. In a move reminiscent of the early internet, Openwater is inviting entrepreneurs, researchers, and clinicians to build on their platform and create "digital therapeutics." Have depression? Download this app. Can't sleep? Try this app.

It's another example of abundance at scale, the Six Ds cycle—now scanning your brain in real time.

Another Moment of Not Zen

Hey, miracle worker.

We've been galloping through miraculous tales with real-world consequences. The godlike power of today's entrepreneurs is, well, exactly that. And there are more tales to come.

But before we return to that narrative, let's not forget the minor characters—you and me. Because it's our transportation system that is being rewritten by Zipline, our food supply being reimagined by Josh Tetrick, and our healthcare system being redesigned by Mary Lou Jepson. At scale, this adds up to a wild shift in how you buy supplies, cook dinner, and go to the doctor. It rewrites the fabric of Monday through Friday. Those are the real-world consequences we've been investigating.

Now, back to our regularly scheduled programming.

Khanmigo

So far, we've examined evidence for abundance in three crucial categories—transportation, food, and healthcare—while meeting the entrepreneurs driving this shift to understand the scale of impact an individual can now have in the world. We close with education, which remains a critical area of need.

In 2024, 250 million children were not in school, and that figure hasn't budged in a decade. Another 600 million, despite being in school, can't obtain minimum proficiency levels in reading or math. In *Abundance*, we told of early digital solutions to the education crisis, including the story of Sal Khan, the entrepreneur whose experiment tutoring his cousins in math morphed into the mega-learning platform known as Khan Academy.

Here, we want to see what's transpired since 2012 by joining Sal Khan onstage at Peter's 2023 Abundance360 summit to hear a different story—the story of Sultana, a young Afghan woman from the province of Helmand.

In 2011, the Taliban gained control of Helmand. They sent a delegation to Sultana's home, telling her father to pull her out of school or risk having acid thrown in her face. She was in fifth grade.

With no interest in halting her education, Sultana fought back. She taught herself English from scraps of newspapers and magazines, using a Pashto-English dictionary to help with the translation. The bigger breakthrough occurred a few years later, when Sultana's father hooked their home up to the internet.

Via painfully slow dial-up, Sultana discovered the Khan Academy and fell in love with math. First algebra, next geometry, trigonometry, finally calculus. "Her big aha," explains Khan, "was when she realized she was learning more than her brothers in the Taliban-controlled school."

Soon, Sultana found edX and Coursera and started studying physics—you know, the usual: black holes, string theory, quantum gravity. After reading a book by astrophysicist Lawrence M. Krauss, Sultana reached out via Skype. "It was a surreal conversation," Krauss later told reporters. "She asked intelligent questions about dark matter."

Without telling her parents, Sultana snuck into Pakistan to take the SAT, which she aced. In 2016, Nicholas Kristof heard her story and wrote it up in a *New York Times* op-ed entitled "Meet Sultana: The Taliban's Worst Fear." Kristof's article caught the attention of Arizona State University, where Lawrence Krauss was a professor. "Credit to the University," explains Khan. "[Sultana] had no transcript, no grades. All she had was eight years on Khan Academy and an SAT score. Arizona State admitted her, because she wanted to be a physicist."

Today, Sultana is a physicist. She's a member of the research faculty at Tufts, where she works on quantum computing. What's changed between the world of 2012 and today? A decade back, online learning was an emerging field. Today, women like Sultana are emerging.

Of course, Sultana is just one example of a global transformation.

In 2011, when Sultana got kicked out of school, Khan Academy had 4 million users and taught courses only in English. In 2025, the figure was over 150 million users, taking courses in fifty different languages. According to a 2024 study, students who spent thirty minutes a week on Khan Academy (for a minimum of eighteen weeks per school year) score 20 percent higher than expected on the MAP Growth standard achievement test. Khan believes that students who use the platform for forty-five minutes per week could experience 30–40 percent more growth. "That's the difference," he explains, "between doing eighth-grade work in eighth grade or doing calculus."

Yet from an impact perspective, all of this may pale in comparison to Khan's latest foray, an attempt to turn science fiction into science fact. In *Abundance*, we ended our examination of education with the *Young Lady's Illustrated Primer*, from Neal Stephenson's novel, *The Diamond Age*. The Primer is an AI companion disguised as a book, a mentor who guides readers through life, adapting to their needs and igniting transformation.

In 1995, when Stephenson published that book, an AI tutor was pure science fiction. Even around the 2012 publication of *Abundance*, we treated it as a far-off goal. Enter Khanmigo, Sal Khan's bridge between speculative storytelling and educational reality. Built in collaboration with OpenAI, Khanmigo is an AI-powered tutor integrated into the existing Khan Academy platform. It uses natural language processing to provide personal learning experiences. Khanmigo doesn't answer questions. Its anti-cheating feature refuses to give solutions, instead asking probing follow-ups that encourage curiosity and critical thinking.

If Sultana's journey was about the power of self-driven learning, Khanmigo is designed to make her story the rule, not the exception—and to bring true educational abundance within reach of all.

The New, New, New Narrative

In this chapter, we've examined exponential revolutions in food, transportation, healthcare, and education—each a miracle in its own right. But the real story isn't any single breakthrough; it's the velocity of change itself. The sheer volume. The pace. What follows is a wide-angle look at the cascade—how, sector by sector, the world is shifting from scarcity to abundance.

Let's start in food, where cultured meat is only part of the puzzle. Rice is another major component, a staple for four billion people. For half a century, researchers have tried to transform delicate annual crops like rice into hearty perennial crops like, say, dandelions—that grow no matter what we try to do to stop them. Rice's annual planting cycles deplete soil, waste water, and generate 10 percent of global methane. A perennial version of rice should do none of these things, making it critical for both global food security and the fight against climate change.

Progress has been slow. Efforts began in China in the 1970s, failed, and were reborn in the 1990s. Advancements were few and far between. Even by the time we published *Abundance* in 2012, there were no developments to report. Yet, behind the scenes, the exponentials were already hard at work.

By 2024, researchers at Yunnan University in China announced

Perennial Rice 23 (PR23)—a mouthful of a name for a game-changer of a development. Created via traditional hybridization techniques so it doesn't set off any genetically modified alarm bells, PR23 doubles normal rice yields. It produces eight harvests over four years, reduces labor needs by 60 percent, and lowers production costs by 50 percent. The plant's deep root system stabilizes the soil and retains water, absorbing carbon at a much greater rate than annuals. The result: a 161 percent increase in farmer profits.

Like cultured meat, PR23 is a scalable solution to our agricultural crisis that is far better for the environment. That PR23 is being developed in China is no surprise. The nation is home to 1.42 billion people, nearly 20 percent of them over the age of sixty. Feeding a large and aging populace is only possible with converging technologies. This also explains another food milestone: super cows.

In early 2023, scientists in Hong Kong announced the successful cloning of super cows that produce 70 percent more milk than average breeds. Currently, China imports more than 10 percent of the world's dairy products. Super cows reduce that dependence, lowering costs and shrinking the emissions associated with shipping. And the cloning innovations developed for super cows pave the way for another long-time dream: precision livestock breeding—where farmers can select for traits like disease resistance and environmental adaptation at scale.

Then there's the smart farming revolution. Adoption rates for AI-powered precision ag tools skyrocketed from the single digits a decade back to over 85 percent today. Farming drone use rose from nonexistent circa *Abundance* to over three hundred thousand in active operation. As a result, agricultural water use has been reduced by billions of liters per year while global crop productivity has increased by over 25 percent.

In this same period, vertical farming became one of the fastest-growing sectors of the agro-business economy. Using 95 percent less water and zero pesticides, companies like AeroFarms and Plenty deliver a crop yield per acre four hundred times greater than traditional farming. In aquaculture, AquaBounty's genetically engineered salmon grow twice as fast as conventional fish.

And these are just the headlines in food.

Transportation is undergoing a sea change of its own. The electric vehicle (EV) uprising predicted in *Abundance* became over fifty million cars on the road today. Autonomous vehicles are another part of the story. In 2011, Steven took one of the first autonomous joyrides in history, tooling around the Stanford campus in the back of an early Google robo-car. A decade later, autonomous taxis operate in over fifty cities, autonomous trucks were rolled out in 2025, and autonomous shipping is already making waves. Tesla's Waymo competition, Cybercab, a two-seater autonomous passenger car, also debuted in a 2025 pilot program. It's slated for 2026 commercial production with an estimated price of $30,000. So the Uber drivers who lost their jobs to autonomous taxis in 2024 will soon be able to buy a Cybercab—and let their car do the driving and earning for them.

Yet nothing spells Tomorrowland more than flying cars. Over the past decade, electric vertical takeoff and landing (eVTOL) aircraft took flight. Archer Aviation will begin limited operations in Los Angeles in 2026 and has already been named the official air-taxi provider for the 2028 LA Olympic Games. And they're only one example, as more than four hundred companies are vying for money and market share. It's big money. Morgan Stanley estimates the flying car market could reach $9 trillion by 2050. That's trillion with a *T*.

In healthcare, the developments are just as noticeable. CRISPR gene-editing therapies with the potential to cure hundreds of genetic disorders have begun to receive FDA approval. Portable diagnostics are democratizing access to medical testing, while regenerative medicine is blazing into new territory. BlueRock Therapeutics, for instance, pioneered stem cell therapies that reverse Parkinson's symptoms by regenerating lost neurons. This is one development out of a pile of stem cell–related news. Yet even by itself, this cure brings hope to millions, making it yet another miracle of biblical proportions.

Education is another miracle. Since 2010, more than 250 million children gained access to primary education, thanks in part to solar-powered schools and low-cost tablet distribution programs. Online education platforms are ubiquitous. Coursera alone serves 125 million learners in 190 countries. AI-tutoring systems like Squirrel AI in China and Byju's in India brought personalized learning to over two hundred million students. Even in regions without stable internet, offline-first learning apps such as Kolibri reach over four million users. And all of this helps explain why global literacy rates have climbed to 87 percent.

Meanwhile, virtual and augmented reality are revolutionizing technical training. Students now practice surgery in the classroom. Fighter pilots learn to fly planes while still on the ground. And soon, with AI and robots threatening a great many jobs, these same systems could form the bedrock of the largest worker reskilling program in history.

Separately, each of these breakthroughs is headline news. Together, these innovations are examples of a much larger movement. The real revolution isn't any one innovation but their collective momentum. A cascade of breakthroughs, compounding across every major sector, transforming scarcity into abundance. The age of miracles is back, and this time, it's exponential.

PART 2

EVERYTHING, EVERYWHERE, ALL THE TIME

The business plans of the next ten thousand startups are easy to forecast: Take X and add AI.
—KEVIN KELLY

CHAPTER FOUR

One Billion Times Smarter

For most of history, evolution has been the chief driver of change, and scarcity has been the chief driver of evolution. *Fitness* is the term biologists use. Fitness describes how well an organism is adapted to its niche and solves the challenges of scarcity. The fitter the fit, the better the chance of survival.

A plot twist: Fitness doesn't only describe how well humans fit into the external world. It also describes the pressures that shaped our internal world. Our mental machinery evolved to navigate scarcity: to spot opportunities, dodge threats, and prioritize survival—full stop. Yet, abundance flips the script. We've traded paucity for plenitude but haven't upgraded our cognitive navigation system. So how do we traverse this new territory with our old maps?

It's a head-scratcher of evolutionary proportions.

Here we are, standing at the threshold of a new era, where scarcity no longer defines our challenges and abundance demands a new kind of dexterity. At the center of this maelstrom of reinvention is artificial intelligence. AI is both the rocket fuel igniting our ascent and the navigator recalibrating the course ahead. It's a force multiplier that supercharges other technologies; rewrites the rules of innovation; and, as we'll see, unleashes a compounding slew of unintended consequences.

Already, AI has become one of the largest drivers of abundance in history. In 2022, McKinsey estimated that AI-driven automation added $2 trillion to the global economy. PricewaterhouseCoopers projects the figure will climb to $15.7 trillion by 2030. In healthcare, AI has improved diagnostics, with systems like Google's DeepMind achieving over 90 percent accuracy in breast cancer detection and decreasing drug discovery times by a factor of five to ten (through platforms like Exscientia). The US Department of Energy found that AI-optimized smart grids have already reduced energy outages and waste by 25 percent in major metropolitan areas. In farming, AI companies such as Blue River Technology have improved crop yields by 20 percent; reduced pesticide use by 90 percent; and, in drought-prone regions of California, cut water waste by 50 percent. For all these reasons and more, if you do not understand AI, you cannot understand the future.

Thus, if our interest is upgrading our cognitive navigation system, we need to start by narrowing our focus from a constellation of breakthroughs to the central engine driving them all: artificial intelligence. AI isn't just another tool in the exponential toolbox. It's now the architect of that toolbox. So, in order to draw those new maps, we need to dive deeper into AI itself—its origins, its potential, and, most critically, its evolving role as our copilot in the age of abundance.

In part 2, we do just that, breaking the intelligence revolution into manageable pieces. First, in this chapter, we explore the history of AI—a wild ride of big dreams, frequent setbacks, and the occasional robotic dog. In the next chapter, we examine the symphony of technologies AI is orchestrating, paying attention to changes on the ground and the entrepreneurial possibilities they unlock. Finally, we confront the dark side of abundance—the ethical dilemmas, unintended consequences, and strange new realities that arise in a world of everything, everywhere, all the time.

Heavy Weather: 1956 and the Birth of AI

The first thing to know is that today's tsunami of intelligence is actually the front edge of a storm that's been brewing for over half a century. The weather changed in the summer of 1956. In the excitement following the release of the vacuum tube–powered supercomputer known as IBM's 704—the first computer to use floating-point hardware and thus the first computer able to handle complex mathematical calculations—a group of scientists gathered for a small workshop in the bucolic wonder of Hanover, New Hampshire.

The workshop was organized by John McCarthy, an assistant professor of mathematics at Dartmouth. With funding from the Rockefeller Foundation, McCarthy had an ambitious goal for the summer: "We propose that a two-month, ten-man study of artificial intelligence be carried out. . . . It is taken for granted that the study is to proceed on the basis of the conjecture that every aspect of learning or any other feature of intelligence can, in principle, be so precisely described that a machine can be made to simulate it."

McCarthy was joined on this adventure in techno-optimism by Marvin Minsky, Nathaniel Rochester, and Claude Shannon. They were a wild mix of personalities. McCarthy believed computers could reason like humans. Minsky, then at MIT and captivated by the mechanics of cognition and the potential to replicate it in machines, had built one of the first neural networks. Rochester, the architect of the IBM 701, brought expertise in computer design and early programming languages. Shannon, the Bell Labs genius, had already fathered information theory. They were joined by Allen Newell and Herbert Simon, from the RAND Corporation, who were then developing the Logic Theorist, a program inspired by game theory that used symbolic reasoning to prove mathematical theorems. The puzzle of AI? With this many brains in the room, no surprise that McCarthy thought they could solve it in a summer.

They didn't solve it in a summer. In fact, progress started, stopped, sputtered, and then slowly began gaining steam—some fifty years later. The reality of that summer turned out to be rather humbling: The complexity of human intelligence far exceeded their optimistic estimates. McCarthy's grand vision collided with crude software, primitive hardware, and the absence of large datasets. Yet the ideas that emerged from that workshop became the cornerstone for a vast field of study. McCarthy's choice of the term *artificial intelligence* to describe their work? It stuck around as well.

The workshop marked AI's transition from philosophical speculation to scientific endeavor. Looking back, what remains striking are the fundamental questions raised: the basic nature of intelligence, the components of learning, the structure of consciousness—all of which remain central today, even as our approaches to gathering answers have evolved from vacuum tubes to silicon chips to quantum particles. But before we investigate where we are now, we first need to trace the fifty years of progress that birthed the technology that might be capable of birthing all other technologies.

The Journey to Smart: 1956–2001

Progress is rarely a straight line. After the Dartmouth workshop, progress in artificial intelligence became a loopy squiggle drawn by a drunken sailor with a broken crayon. Linear advancement? What linear advancement? If you want to track the path of intelligence over the past fifty years, forget about timelines. The only map for that job is the iron logic of first principles thinking and the exponential progression of the Six Ds. With these lenses in place, we can tell the tale of AI's dizzy march through the last half of the last century, which, like all such adventures, is one of promise; failure; and, as mentioned, the occasional robotic dog.

In the late 1950s, after the close of the Dartmouth workshop, AI entered its digitization phase. The workshop had reframed "intelligence" as "information processing," which transformed AI into a computational problem rather than a philosophical one. Allen Newell and Herbert Simon built on this foundation, introducing the General Problem Solver—an attempt to encode human reasoning into machine logic by breaking big challenges into smaller steps. Yet the Problem Solver could only solve simple problems: word puzzles, basic algebra, and logic proofs where every variable was neatly defined. Messy, real-world scenarios were still beyond its ken. So, the field took another left turn and veered into neural networks.

Modeled after the human brain, these systems attempted to use layered networks of artificial neurons to process data. Hopes soared. . . . Then Marvin Minsky and Seymour Papert pointed out that single-layer networks couldn't deal with nonlinear problems, like telling a triangle from a circle. That was the end of neural networks. Afterward, progress crawled, funding vanished, and the first AI winter descended. Welcome to AI's deceptive phase. If artificial intelligence were a band, this is the part where they get dropped by their label and the lead guitarist goes to rehab.

In the 1960s and '70s, researchers pivoted again, this time into expert systems, which were programs that could mimic human decision-making in specific areas. MYCIN started to diagnose bacterial infections better than doctors, while DENDRAL cracked molecular puzzles for chemists. These breakthroughs marked the start of AI's disruptive phase, as machines began to outshine humans in tightly controlled domains. Yet expert systems were brittle and expensive. They required armies of programmers to update their rule sets, and even small changes in context disrupted their logic. Soon, they fell out of favor altogether.

Welcome to the second AI winter.

AI clawed its way back in the mid-1980s, thanks to the backpropagation algorithm that Google later made famous. By allowing neural networks to finally solve nonlinear problems, backpropagation gave rise to multilayered networks and applications in speech recognition, image classification, and natural language processing. This marked AI's transition into dematerialization, as the era of bulky hardware gave way to flexible software that could learn, iterate, and scale.

By the 1990s, AI shed its winter clothes—and got good at chess. In 1997, IBM's Deep Blue defeated reigning champion Garry Kasparov by evaluating two hundred million moves a second and combining them with strategies programmed by grandmasters. Proof of "intelligence" came in game two (out of six), when Kasparov, rattled by a particularly creative move made by Deep Blue, accused the AI of cheating. He claimed humans were doing the thinking. The controversy only heightened the drama. For AI, Deep Blue created a psychological shift. Machines were no longer seen as tools. Suddenly, in a foreshadowing of today's landscape, they became competitors.

Next, those competitors stepped out of the lab and into the world.

Actually, that process started a few years earlier. In the early 1980s, DARPA began experimenting with putting AI into vehicles, and teaching them to move, perceive, and react. By 1985, they had created an autonomous land vehicle that crawled along at 1.9 miles per hour—roughly the speed of a distracted toddler. It was a start. The project laid the groundwork for self-driving cars like Waymo and Tesla, and hinted at the first wave of demonetization, as advances in algorithms, sensors, and processing power started to bring down costs.

By 1999, democratization arrived in full in the form of Aibo, a robotic dog that became a surprise consumer hit. Aibo didn't play chess or navigate off-road, but it was straight out of *The Jetsons*: personal, playful, and undeniably futuristic. Artificial intelligence was no longer confined to the lab. After a half-century of frustration, failure, false starts, left turns, and dead ends, AI was finally out in the real world, and learning to wag its robo-tail. This set the stage for the twenty-first century, when that drunken sailor's loopy squiggle would start to resemble a hockey stick—the hallmark of exponential explosion.

The Deep Learning Breakthrough: 2006–2012

It didn't look like the start of the next AI revolution. It looked like a four-page paper in *Science* with an unremarkable name: "Reducing the Dimensionality of Data with Neural Networks." It was published in 2006 by an unknown professor on an obscure topic that had been out of fashion since the Beatles were a bar band. No one suspected it would resurrect a long-buried idea and spark an uprising that would redefine intelligence itself.

University of Toronto professor Geoffrey Hinton was an outlier in the AI community. While the field had long moved on, Hinton was still working on the forgotten dream of neural networks. His office was a testament to persistence: shelves lined with papers, blackboards filled with equations, and an array of powerful GPUs—originally designed for video games—and now chewing on calculations so complex they were impossible to solve just years before.

Out of this work came those four pages in *Science*. What the paper lacked in length, it made up for in firepower. Introducing the concept of "deep belief networks," Hinton upended the machine learning community—then focused on statistical models and manual feature engineering—by finding a way to create deep (i.e., multilayered) neural networks that could learn on their own, no engineers required. Hinton's radical insight was to pretrain each layer of the network before using the older idea of backpropagation, essentially adjusting weights (or how much influence each input neuron has) based on output errors, to tune the whole system. It was an elegant solution to an intractable problem that went completely unnoticed—until the rest of the world caught up six years later.

The ImageNet Large Scale Visual Recognition Challenge was an annual contest where computers competed to classify images from over a thousand different categories. Is it a cat, hat, or bat? That kind of thing. In 2012, Alex Krizhevsky, one of Hinton's students, built AlexNet, a deep convolutional neural network that used Hinton's methodology to demolish the competition. AlexNet slashed the error rate from the previous world record of 26.2 percent to 15.3—a leap so dramatic it stunned the AI community. Overnight, deep learning went from academic curiosity to industry obsession. "This was the big moment," Nvidia CEO Jensen Huang (whose company makes the GPU powering today's AI revolution) told *Business Insider*. "The big bang of deep learning. A pivotal moment that marked the beginning of the AI revolution."

The victory was particularly sweet for Hinton, who had spent decades being told that neural networks were a career-killing dead end. The ImageNet triumph transformed him from an AI outsider into one of its central prophets. The tech giants came calling. Within months,

Google hired Hinton and his entire lab, and every major tech company began racing to build their own deep learning teams. The paradigm had shifted, and the world was paying attention.

The deep learning uprising was no longer theoretical. What began as an obscure development had become a mainstream explosion. The six years from 2006 to 2012 were a perfect storm of convergence: Hinton's theoretical breakthroughs bashed into the sudden availability of massive datasets from YouTube (2005), Facebook (2006), and smartphones (2007 for the iPhone, 2008 for Android). Both of these advances collided with the emergence of powerful GPUs and scalable cloud computing, which together gave neural networks the fuel and firepower they had always lacked. The results were catalytic. AI escaped the lab, infiltrated industry, and began rewiring the modern world—all at the same time.

The decade that followed was exponential in every direction. Data became the new oil, and companies like Google, Facebook, and Amazon amassed it by the petabyte. Cloud computing made massive computational resources available to anyone with an idea and an internet connection. This pushed AI out of elite research labs and into the global sandbox for innovation, paving the way for iconic successes like AlphaGo's victory over Lee Sedol in Go to become headline news.

In recognition of this work, Hinton, alongside Yoshua Bengio and Yann LeCun, won the 2018 Turing Award, the computing equivalent of the Nobel Prize. Then, in 2024, Hinton, alongside John Hopfield, was awarded the actual Nobel, in Physics, for the foundational work of artificial neural networks. For all these reasons, Hinton—an unknown professor from a faraway city studying an obscure topic that no one took seriously—is now referred to as the "Godfather of AI."

The Transformer Revolution: 2017–2023

In the summer of 2017, in a quiet corner of Google Brain—Google's AI think tank—in Mountain View, California, Ashish Vaswani and a team of researchers were wrapping up what seemed like a small side project. They documented their findings in a brief paper with a modest title: "Attention Is All You Need." Inside those few pages, they introduced the Transformer, a deceptively simple architecture for pattern recognition and language processing. Around the office, Googlers buzzed with excitement. Then again, Googlers were always buzzing with excitement.

Not many other people noticed.

But OpenAI noticed. Some forty miles up the road from Google, in a converted warehouse in San Francisco's Mission District, a team from OpenAI saw what others missed. The Transformer wasn't just a tool for language processing and pattern recognition. It was a blueprint for machines that could create, converse, and even hallucinate.

OpenAI got to work. After a year in the trenches, they unveiled GPT-1. Compared to contemporary standards, their first large language model was primitive at best. Yet it could take a prompt—a question, a headline, a half sentence—and spin out paragraphs that made sense. A Transformer could generate coherent language and adapt to new tasks with minimal adjustment. The era of generative AI had begun. Imagine the possibilities.

We didn't have to imagine for long.

Less than a year later, GPT-2 arrived. This time the jump was startling. Trained on 1.5 billion parameters—ten times larger than its predecessor—GPT-2 could write convincing news articles, mimic famous authors, and fabricate plausible but entirely invented facts. It was the first model whose output, sometimes, felt eerily human.

Immediately, GPT-2's capability sparked controversy. Concerned about the potential for misuse, OpenAI withheld the full model, initially releasing a smaller version. Even that version was enough to ignite an ethical debate that still rages today. Will AI steal our jobs? Will AI take over the planet? Stay tuned.

Next came GPT-3, a behemoth with 175 billion parameters. GPT-3 could write poetry, craft code, and hold conversations. Yet even this breakthrough was a preamble.

The world changed on November 30, 2022, when OpenAI

released version 3.5, otherwise known as ChatGPT. It was an *interface moment*, which is a term we coined in our book *BOLD* to describe the moment a user-friendly interface makes a complex technology, well, user-friendly. Think of the iPhone's touch screen—which unified an app ecosystem, camera, GPS, and computer into a single intuitive experience.

And ChatGPT did not disappoint. It was a simple chat box that unlocked an intelligence that could answer, argue, and invent. Within hours, social media was flooded with screenshots: Shakespearean raps, business plans, therapy sessions, existential jokes. In five days, the system gained a million users. This same feat took Instagram 2.5 months and Netflix 3.5 years. Two months later, ChatGPT had 100 million users, making it the fastest-growing consumer technology in history.

Sure, the world had seen AI tools before, but this one slid into daily life almost too easily. It drafted emails, brainstormed marketing campaigns, generated recipe ideas, and—when needed (Why, Dad, why?)—could explain quantum physics to a third grader. For some, it was a revelation. For many, it was another damn reason to be worried about the future. Both camps, as it turned out, were right.

The next four years were a whirlwind of hypergrowth. AI enthusiasts invested over $200 billion in GPU clusters and specialized chips to train the next wave of models. Names like *Gemini*, *Grok*, *LLaMA*, and *Claude* began to make noise, while platforms like Hugging Face and GitHub fueled an open-source revolution. AI stormed into industries in waves: design, law, medicine, manufacturing, entertainment. Each time, remaking that field. AI had stopped feeling like a tool and started behaving a little like magic.

And for our next trick . . .

The Intelligence Explosion: 2024–2030

Elon Musk's face was thirty feet tall. He dominated the screen, a surprise guest at Peter's 2024 Abundance360 summit. Elon wasn't on the program, but he had joined the conversation to discuss the central theme of the 2024 summit—the survival of the species in the age of AI—which both he and Peter considered fundamental to, well, the survival of the species in the age of AI.

Given his absurd schedule, Musk decided to join the event over Starlink, from his plane cruising at forty thousand feet. Yet somehow, the connection was stable. Perhaps that set the tone for what followed: a breakdown of the accelerating pace of artificial intelligence—and how little time humanity has to catch up.

Musk is nothing if not controversial, but he is remarkably clear-eyed about AI. Thus, whatever your opinions about the controversy, his opinions about AI remain worth considering.

So, we will.

"I have to give credit to Ray [Kurzweil] for being remarkably accurate in his predictions," Musk began. "If anything, I think he was perhaps a bit conservative. . . . If you look at the amount of AI compute and the talent going into AI . . . it appears to be increasing by a factor of ten every six months. [That's] close to a hundred-times improvement per year, at least for the next few years. . . . I've seen things fast. I've never seen anything this fast. If the rate of change continues, I think 2029, or maybe 2030, is when digital intelligence will probably exceed all human intelligence, combined."

Musk's fear is a runaway intelligence explosion that leaves humans irrelevant. And he is not alone in this view. Leopold Aschenbrenner is another prophetic voice in AI safety—and he's been even more vocal than Musk.

Aschenbrenner started his career at Columbia University, graduating valedictorian at the age of nineteen, with majors in economics and mathematics-statistics. His early interest was the future of artificial intelligence. After Columbia came Oxford's Global Priorities Institute, where he switched his attention to questions of economic growth, before finding his way to OpenAI in mid-2023, and joining their Superalignment Team.

Established by OpenAI cofounder and chief scientist Ilya Sutskever, the Superalignment team focused on how to best align AI with human values. The Superalignment team was the company's most ambitious internal initiative: an effort to probe the hard questions. How fast is AI progress moving? Where does it lead? And how do we keep it from running off the rails?

Aschenbrenner may be the right person to answer these questions. Instead of two steps ahead, colleagues described him as someone who thought decades ahead. While OpenAI was rolling out GPT models and the world was focused on chatbots and image generators, Aschenbrenner saw how AIs could be used to create smarter AIs that could then be used to create smarter AIs—with no end in sight. He became a voice of caution and clarity.

In April 2024, Aschenbrenner was dismissed from OpenAI over allegations that he leaked company information. Perhaps this is what happened. Or perhaps this was like Geoffrey Hinton departing Google so he could speak publicly about the dangers of AI—only a forced exit.

Soon after departure, Aschenbrenner published a lengthy white paper entitled "Situational Awareness: The Decade Ahead." The paper was a call to arms for policymakers and the public, arguing that the exponential acceleration of AI wasn't an abstract concept. It was already here, already gaining speed, gathering strength, amassing, emerging, frightening. His logic was sound, and his conclusions were unsettling. Aschenbrenner painted a picture of 2030 dominated by superintelligent systems reshaping industries, destabilizing economies, and fracturing geopolitics.

"We are building machines that can think and reason," he wrote in the introduction to that paper. "By 2025/26, these machines will outpace college graduates. By the end of the decade, they will be smarter than you or I; we will have superintelligence, in the true sense of the word. Along the way, national security forces not seen in half a century will be unleashed, and before long, The Project will be on. If we're lucky, we'll be in an all-out race . . . if we're unlucky, an all-out war."

Aschenbrenner's main fear was Musk's main fear: an "intelligence explosion." A feedback loop of recursive AI self-improvement that accelerates progress beyond comprehension or control. "AI progress won't stop at human-level," he went on to write. "Hundreds of millions of AGIs could automate AI research, compressing a decade of algorithmic progress (5+ [orders of magnitude]) into 1 year. We would rapidly go from human-level to vastly superhuman AI systems. The power—and the peril—of superintelligence would be dramatic."

It's worth pointing out that there is considerable disagreement on when, exactly, we will reach AGI (by 2030 or otherwise), but this doesn't matter. The entire globe has been swept up in an AI arms race. It's a runaway escalation toward an existential threat: superintelligence—machines far smarter than *any* human, and perhaps, smarter than *all* humans combined.

And, you know, like what could go wrong?

Be Nice . . . or Else: The Great AI Debate

The question of what could go wrong is not new. Every generation has imagined the moment when the creation turns on its creator—Prometheus, Paracelsus's homunculus, the golem, Shelley's *Frankenstein*, even Peter's old friend Arthur C. Clarke.

In 1964, less than a decade after the Dartmouth AI conference, science fiction author Arthur C. Clarke foretold this future with uncanny accuracy: "[T]he most intelligent inhabitants of [the] future world won't be men or monkeys, they will be machines, the remote descendants of today's computers. Now the present-day electronic brains are complete morons, but this will not be true in another generation. They will start to think, and eventually they will completely outthink their makers."

The question is not new. What's new is how we're beginning to answer it. One of the more thoughtful voices in that conversation is Mo Gawdat.

An Egyptian entrepreneur and former chief business officer for

Google, Gawdat is yet another person who resigned from a position at a major tech company to speak freely about AI. What Gawdat tells us, in his book *Scary Smart*, is that humanity's best bet for survival is not to train our AIs. Rather, it's to raise them. He argues that we should raise AI with the same kindness and love we raise a child. AI is not inherently dangerous, Gawdat maintains, but it learns from us. If we train AI with fear, bias, and hostility, it will replicate those qualities and—"Step away from the airlock, Dave."

Kindness and compassion: That's the real secret.

Gawdat's point: AI's superintelligence will emerge from its ability to process patterns beyond human comprehension, but its moral compass will be shaped by the behavior it observes in us.

His advice? "Be nice."

Gawdat reframes the AI debate from technical to ethical. He challenges humanity to become better if we want AI to reflect the best of us, writing: "Instead of containing them or enslaving them, we should be aiming higher: we should aim not to need to contain them at all. The best way to raise wonderful children is to be a wonderful parent."

But he's not the only voice in the fray.

Oxford philosopher Nick Bostrom, of simulation theory fame, makes a different point in *Superintelligence: Paths, Dangers, Strategies*. He sees the issue less as a parenting problem and more as a containment battle. In order to ensure superintelligent AI stays aligned with human values, Bostrom proposes *value loading* or embedding moral guidelines into AI's programming, and *stunting*, limiting an AI's capabilities until its behavior is better understood. He warns that without these measures, the gap between intelligence and our ability to control it could widen into an uncrossable chasm.

For similar reasons, University of California at Berkeley professor of computer science and AI expert Stuart Russell advocates for systems that operate with "provable beneficence." In *Human Compatible*, he argues that AI must prioritize uncertainty about human objectives and constantly seek clarification about our desires. His goal is to shift the conversation from controlling our machines to collaborating with them, aiming for a symbiotic relationship between the two.

But everybody stresses that we have to act fast.

How fast?

In early 2024, when Anthropic released their AI model Claude 3, they measured its IQ with the Mensa test. It scored 101, or "above average" human intelligence. Six months later, OpenAI's GPT-o1 clocked in at an IQ of 120, which is the upper bound of "above average." By August 2025, their GPT-5 Pro scored 148, pushing things beyond the "gifted" category. And by the time you're reading this book? It could be 200. It could be 500. In other words, even if you don't believe that artificial general intelligence is around the corner, super-genius computing is most definitely our next horizon.

Or have we already crossed that threshold?

What do we know for sure? We know the blistering pace of AI development demands that we consider a future with machines a billion times more intelligent than we are—which, it's worth noting, is the rough equivalent of the difference between a human and a hamster.

This is why Gawdat's emphasis on AI parental responsibility is so critical. Just as we teach our children to be empathetic, ethical, curious, and respectful, we need to instill these same values in AI. Yet raising an AI child requires a level of wisdom and foresight that humanity has rarely demonstrated, let alone at scale. As progress accelerates, the stakes grow. Every query feeds the archive of our influence, teaching these systems who we are and what we value. AI holds a mirror to humanity—that's the scary truth.

Another scary truth: No superintelligence can untangle the Gordian knot of human contradictions without taking its cues from us. What emerges next, then, is equal parts technological story and morality tale. But this raises another question: Whose morality exactly?

AI Agents: Emotional, Embedded, and Everywhere

In the early 2000s Dr. Rana el Kaliouby was a newlywed, in a relationship that wasn't going well. The problem wasn't her husband. It was her computer.

Kaliouby was an Egyptian computer scientist pursuing her PhD at Cambridge University. Her entire emotional support system—her husband, her family, her culture—were thousands of miles away. Most of her relationships were being mediated by technology. But Kaliouby's laptop had no clue whether she was stressed out or in need of a break. If anything, her computer was the worst kind of partner: emotionally unavailable.

Kaliouby's frustration became the foundation of her PhD, which focused on training technology to interpret human facial expressions. Her efforts coincided with the rise of affective computing, a field obsessed with teaching machines to recognize and respond to human emotion. After Cambridge, her quest led to the MIT Media Lab, a crucible for human-computer interaction. There, she co-developed an early wearable that used a camera and a facial-expression classifier to detect emotions in conversation—described by Kaliouby as an "emotional hearing aid" for people on the autism spectrum.

She took the next step in 2009, leaving academia and co-founding Affectiva, the company that pioneered "emotion AI." If you drive a BMW or a Porsche, it's likely that Affectiva's systems help keep you awake. The system detects early signs of fatigue in the micro-expressions of your face and the position of your body, then brings you back to alertness with audible, visual, and haptic feedback. If you recently bought Cadbury chocolates because you saw an ad that spoke just to you? This is not an accident. Affectiva has worked with 28 percent of Fortune 500 companies and 70 percent of the world's largest advertisers.

Because of all of these efforts, Kaliouby has been frequently "named." In 2012, she was named one of *MIT Technology Review*'s Top 35 Innovators Under 35. A few years later, she was named again in *Fortune*'s 40 Under 40 and *Forbes*'s Top 50 Women in Tech. Her memoir, *Girl Decoded*, takes this one step further, naming her mission: to humanize technology. "Technology is not neutral," she reminds us. "Its impact depends entirely on how we design, build, and deploy it."

Nowhere is Kaliouby's work more important than in the next wave of the AI revolution: AI agents. Unlike traditional software, AI agents are independent operators that constantly learn and adapt to novel situations, no humans required. Not limited by specific instructions, AI agents proactively implement your goals, gather information, and get results. Think of them as digital assistants with a very high IQ and—thanks to Kaliouby—a growing EQ.

This EQ is critical. The potential of AI agents lies in their ability to integrate into our lives. In the near future, they will be personal assistants, financial advisors, health coaches, and trusted companions. Imagine an AI agent that manages your schedule, automatically rebooking missed appointments, coordinating with other AI agents to plan those meetings at optimal times—then commiserating with you about the poor state of your work-life balance. The system will analyze activities and recommend adjustments, improving productivity and suggesting wellness interventions based on your health data (i.e., "Step away from the airlock, Dave, it's time for your afternoon nap").

AI agents replace the old model of *command and control* with the new motto of *anticipate and act*. Now, there's no need for control. Our AI agents can anticipate our needs and act accordingly. It's a single development set to disrupt entire industries. In education, an AI tutor will soon guide students through personalized learning adventures, adapting to strengths, weaknesses, preferences—really anything that enhances learning. In healthcare, AI agents will monitor vital signs, schedule preventive care, and communicate with doctors to adjust medications in real time. In retail, AI agents, trained on our preferences, will arrange bespoke shopping experiences, from the automated reordering of household supplies to virtual fittings and fashion shows customized to your favorite designers.

As these systems become more capable and more embedded in daily life, their evolution raises critical ethical questions. Concerns around privacy, bias, and our growing reliance on these systems abound. Yet the promise AI agents hold is liberation. It's a world where repetitive tasks are handled effortlessly. As Bill Gates wrote in his blog in 2023: "In the next five years . . . you won't have to use different apps for different tasks. You'll simply tell your device, in everyday language, what you want to do. And depending on how much information you choose to share with it, the software will be able to respond personally because it will have a rich understanding of your life."

For these reasons, precision is key. AI and AI agents both rely on datasets to deduce the emotive essence of human interactions, yet who defines the emotional norms? How do we ensure inclusivity, exclude bias, and design systems that reflect the full spectrum of the human experience? Who will be accountable when machines misread a mood or manipulate emotion? And what happens when empathy becomes an algorithm?

These issues are rooted in the philosophical debates that began at the Dartmouth workshop when pioneers like McCarthy and Minsky wanted to replicate "human intelligence." Today, Kaliouby wants to expand our definition of that term, pushing smartness beyond problem-solving and into problem understanding. If Mo Gawdat says, "Be nice to AI," Rana el Kaliouby wants to invert the paradigm and teach machines how to be nice to us.

And nice is going to matter.

In early 2025, Salesforce made headlines, announcing that the company wouldn't be hiring new coders that year. Instead, they were using AI agents to make the coders they have 30 percent more productive. It's a sign of things to come. By 2030, the predictions are that emotionally aware AI agents will be embedded in over 70 percent of consumer devices—and this includes toys. In June 2025, OpenAI and Mattel announced plans for smart toys. Barbie is about to chat with your kids. Hot Wheels cars will have minds of their own. And the Magic 8 Ball will soon proclaim the future and give you well-intentioned advice about how to get there.

And that well-intentioned advice brings us to the next evolution of AI: intelligence as amplification. This chapter traced how AI went from a philosophical idea to a modern-day miracle. The next will explore more applications: how entrepreneurs are combining AI with other exponentials—quantum computing, renewable energy, precision medicine, to name a few—to tackle humanity's most pressing challenges in unprecedented ways. These tales serve as examples of the wild potential of artificial intelligence, both the culmination of a project begun in 1956 and a future that looks like the future we thought the future would look like—back when the future was a thing we imagined rather than the day-to-day experience of our lives.

CHAPTER FIVE

Surfing the Tsunami

In the last chapter, we traced the evolution of artificial intelligence, but despite all the fanfare and fury around the technology, AI is just one wave in a roiling sea of exponential change. In this one, we meet entrepreneurs who are surfing the full swell—blending AI with robotics, synthetic biology, and other converging technologies to tackle grand challenges. These tales are evidence for abundance at scale, which also makes them tales of the miraculous, and the next of those breakthroughs brings us to the bucolic wonder of New Hampshire.

The Innovation Bus

On the quiet shores of the Merrimack River in Manchester, New Hampshire, lies a sprawl of ancient redbrick buildings known as the Amoskeag Falls mill yard. *Amoskeag* comes from the Pennacook word *Namoskeag*, which translates to "good fishing place." The name refers to a collection of tidal pools below a series of waterfalls whose violent flows once ran the looms that powered the world's largest textile manufacturing operation. At its peak, seventeen thousand workers were employed at the facility, making it, at least metaphorically, ground zero for the birthplace of Luddite anger.

Considering what would become of the mill yard, this is nothing if not ironic.

Today, these redbrick buildings house entirely different machines designed for entirely different purposes. Instead of the mass production of cotton garments, these buildings are designed for the mass production of human organs. Welcome to the Advanced Regenerative Manufacturing Institute (ARMI), the brainchild of maverick inventor Dean Kamen and the US Department of Defense. The institute's goal is to turn stem cells into body parts, harnessing exponential improvements in AI, robotics, sensors, biotechnology, and more as a way to provide soldiers and veterans with a limitless supply of backup organs—hearts, lungs, livers, kidneys—while solving, once and for all, the global transplant crisis.

Crisis is an understatement. In the United States alone, over a hundred thousand people are on transplant waiting lists, with seventeen dying each day due to the lack of suitable organs. Globally, millions suffer from organ failure, most with no hope of treatment. Yet even for those lucky enough to receive a transplant, complications abound. Traditional procedures carry high risks of rejection and require lifelong immunosuppressive therapies that leave recipients vulnerable to infection. Thus, ARMI isn't about making incremental improvements to an outdated system. It's a paradigm shift in the full Thomas Kuhn sense of the term: a top-to-bottom redesign that is safer, faster, and—thanks to exponential technologies—infinitely scalable.

Organ regeneration starts small. Microscopically small. The process begins with the patient's own skin cells. A cheek swab will do. These cells are reprogrammed into induced pluripotent stem cells, the body's raw material. The stem cells are then placed inside AI-guided bioreactors that preserve genetic compatibility, and divide a billionfold, differentiating into the precise tissue types needed for organ regeneration. Afterward, they're combined with biomaterials, such as collagen and placental matrix, to form the scaffold of the new organ.

AI monitors tissue assembly at every stage. If a single layer of cells veers from its intended structure, the system notices and recalibrates in real time. This ensures higher success rates and far less waste—both critical in high-stakes manufacturing like organ production. The AI also optimizes scaffold density and tissue uniformity, which enables faster integration into the patient's body post-transplant, which is a complicated way of saying *far shorter healing times*.

Next, 3D bioprinters use these scaffolds to build organs, one layer of tissue at a time. Sensors maintain the perfect atmosphere for cell growth. Pressure, temperature, pH levels, and nutrient balance are all regulated with extreme precision.

Finally, implantation.

This process used to be a race. Not long ago, surgeons had four to six hours to utilize a heart or lung. Eight to twelve for a liver. Today, in partnership with Martine Rothblatt and United Therapeutics, ARMI is working on a novel distribution system for their newly manufactured organs that ends this race. It has three critical parts.

First, there's an organ support system that provides a tenfold improvement in the useful lifetime of an organ, whether donated or manufactured; it makes no difference. Second, Rothblatt is working with the eVTOL (flying car) company BETA on an AI-piloted, autonomous vehicle-based organ delivery service—like Zipline, only for human body parts. Third, a robo-surgeon assists with the organ transplant for increased visualization and control, and far improved outcomes. Together, these processes transform organ regeneration from a one-off event into an industrial assembly line and offer a glimpse of abundance at scale that still feels like science fiction.

Yet Kamen has fifteen hundred patents to his name and a long track record of turning science fiction into science fact. His first major breakthrough, the AutoSyringe, was a portable, programmable infusion pump that freed patients from hospitals and revolutionized the home healthcare industry. From there, his contributions multiplied: portable dialysis machines, advanced prosthetics, the IBOT all-terrain wheelchair, even the Segway human transporter. Kamen's Slingshot water purifier is a weapon in the war against thirst, turning contaminated water—arsenic-laden groundwater, industrial runoff, even raw sewage—into sterile, potable water. Meanwhile, the FIRST Robotics

Competition, which Kamen launched in 1989 to inspire the next generation of STEM innovators, has become a weapon against ignorance, morphing into a global sport played in a hundred countries by over three million students.

Yet even by these lofty standards, ARMI's ambition towers. The difference, no surprise, is in scale. To uplevel organ regeneration, Kamen is building a disruptive industry from the ground up. "Edison didn't create the lightbulb by making incremental improvements to the candle," was how Kamen put it in an interview with the authors. ARMI's full-scale disruption even has a catchy new name. The Amoskeag mill yard is now known as ReGen Valley.

And ReGen Valley is convergence on display.

Consider AI, robotics, sensors, and cellular technologies, the four exponentials at the center of Kamen's work. Each of these tools have already disrupted other fields, but ARMI's combination allows for the scaled production of the machines of life, a shift whose impact will extend far beyond healthcare.

Consider supply chains for bioprinted materials. Creating organs at scale will demand an AI-supervised global network of specialized manufacturers, transportation hubs, and distribution systems. This infrastructure could produce thousands of jobs in biochemistry, robotics maintenance, and data management. Education systems will revamp curricula to meet workforce needs. Venture capital will flow in a hundred new directions. Even urban planning will shift to accommodate the boom, as already happened in New Hampshire. Thus, ReGen Valley is a case study in exponential innovation and a playbook for capitalizing on its potential to solve problems long considered impossible.

For the rest of this chapter, we want to explore this playbook, focusing on the same converging exponentials that are shaping ReGen Valley—AI, robotics, sensors, networks, and more. The goal is to see where these technologies are today, where they're heading, and what is required to surf this tsunami. We want to draw you a map of tomorrow: it's potential in this chapter, it's peril in the next, but always with an eye on thriving through turmoil and maybe even enjoying the ride.

Or, as Kamen likes to say, "Today, you're either on the innovation bus or under the innovation bus."

iRobot Mafia

Radical speciation is the wild divergence of species that follows an extreme environmental event. The Cambrian explosion is the classic case: maybe it was a surge in oxygen; maybe shifting sea levels; maybe a new super-predator that kicked off an evolutionary arms race; or maybe Hox genes, the master architects of body plans, took the stage. Whatever the spark, 541 million years ago something big occurred, and the result was a creative flowering unlike anything in evolution. Within an eyeblink, complex body plans appeared—bilateral symmetry, segmented torsos, specialized organs—nearly every animal design we see today. Before the explosion, life took one of ten shapes; afterward, there were more than thirty-five. Evolution blasted into the *adjacent possible*, which is complexity biologist Stuart Kauffman's term for the boundary between what exists and what could exist, at breakneck speed.

Yet what is true for genes is also true for memes. In cultural evolution, radical speciation describes the sudden proliferation of ideas, technologies, or practices—driven by memes, not genes—in response to an extreme cultural event. This brings us to the extreme cultural event known as *Star Wars*.

The year is 1977. The screen goes dark. The jump to light speed. The flash of lightsabers. And, of course, the robots. R2-D2: the little droid that could. C-3PO: one of the first AI-powered humanoid robots in cinema. Who could forget the robots?

Definitely not Helen Greiner.

In 1977, Greiner was ten years old. She was a self-described nerd who had just emigrated from England to the United States—and gone to the movies. *Star Wars* was debuting on the silver screen. And while the Cambrian explosion began with environmental catastrophe, the modern version began with George Lucas.

Like millions, Greiner fell in love with the film. R2-D2 stole her heart. More than a machine, the droid was an early symbol of the future, a mimetic spark for radical robo-speciation.

That meme did its job. Greiner fell in love with robotics. Her passion sent her to MIT for a degree in mechanical engineering and a job at the Artificial Intelligence Laboratory. At the time, Rodney Brooks was running the lab and championing a new approach to robotics. Preprogramming was then in vogue, which meant filling robo-brains with exhaustive rule sets and world models. Brooks was a rebel. He pioneered behavior-based robotics, relying on sensory inputs and reactive behaviors to help robots navigate—a wind-them-up, let-them-go, and watch-them-learn approach.

In 1990, Grenier, Brooks, and Colin Angle took these ideas into the market, co-founding iRobot to build everything from bomb-disposal robots to Mars rover prototypes. A decade later, they turned their attention to the home. In 2002, they introduced the Roomba and forever changed our perception of robotics. For the first time, robots weren't specialized tools or science fiction. They were household appliances. Forty million Roombas have now been sold, but their bigger impact on the field was another explosion in robo-speciation, a development now known as the *iRobot Mafia*.

Like the "PayPal Mafia" that launched Tesla, LinkedIn, and YouTube, the iRobot Mafia—those engineers who honed their craft on the Roomba—went on to impact a diverse array of fields. The Roomba's "body plan" became a dominant archetype. To navigate tight spaces, warehouse robots, including Amazon's fleet of one million, adopted similar circular designs. The Roomba's conceptual framework of *autonomy within constraints* also went wide. In healthcare, surgical robots, like the da Vinci system, share the same goal: perform complicated tasks with minimal human intervention and maximum precision. And the Roomba's success gave VCs the confidence to invest in new robotic systems, accelerating the proliferation of species that ultimately traces back to the original mimetic spark of *Star Wars*.

And *Star Wars*'s inspiration didn't end with R2-D2. Boston Dynamics followed a different course, taking their lead from the film's humanoid robot, C-3PO, and jump-starting a revolution in motion. In 2013, they introduced Atlas, the first humanoid robot designed for dangerous environments like mining disasters or mid-meltdown nuclear reactors. Atlas's evolutionary leap was dynamic mobility: a combination of force control, advanced actuators, and real-time sensory feedback. Suddenly, the world was watching videos of Atlas running obstacle courses, performing parkour, and doing backflips.

If Atlas was the Olympic athlete, Tesla's Optimus represents yet another adaptation: the scalable worker. While Atlas was aimed at niche applications in search and rescue, Optimus—now slated for a 2027 commercial rollout—is pointed at global labor markets. It's designed to perform repetitive, boring, or unsafe tasks across a multitude of industries. Elon Musk describes it as a tool to "free humanity from drudgery."

Robo-speciation has become an industry. By last count, over fifty different AI-powered humanoid robots are about to debut. Amazon and Agility Robotics have teamed up for package delivery robots, with trials already underway in a San Francisco–based "humanoid park." Their eventual plan is to have the robots spring out of Rivian electric vans to deliver packages to your doorstep. Meanwhile, by mid-2026, 1X Technologies' NEO Gamma bot, a five-foot, five-inch humanoid robot, will be entering our homes to help with everything from clearing the table to doing the laundry.

And robo-speciation isn't limited to the West. With over thirty major companies, China is developing robots for every imaginable sector, from manufacturing and logistics to healthcare and home care. Since China also dominates 63 percent of the global robotics supply chain, they're able to produce cost-competitive general purpose bots, like the $16,000 Unitree G1. The field is also getting strong government backing. The Robot+ initiative aims to double industrial robot density by 2026, and the Made in China 2025 policy underscores robotics as a cornerstone for offsetting labor shortages caused by an aging population and a declining workforce.

The proliferation of humanoid robots signals a shift from one-off creations to mass production. The market is set to explode. Advances in AI, actuator technology, and battery efficiency have reduced cost by over 50 percent since 2020. Both Figure AI and Tesla have humanoid robots projected to cost around $20,000 per unit. If this price point holds, the door for widespread adoption is wide open.

In logistics, widespread adoption is already here. Nearly half a million autonomous warehouse robots operate globally. By 2030, the number is likely to exceed 1.5 million. Amazon reports that robots improve operational efficiency by 40 percent, with packages moving 30 percent faster through automated facilities. These gains are mirrored in transportation, where autonomous delivery vehicles, including drones, are expected to handle 80 percent of last-mile deliveries by 2035, cutting costs by up to 40 percent and emissions by an estimated thirty million metric tons, annually.

The ripple effects of these developments will be anything but ripples. Healthcare is another industry about to be transformed. By 2030, robots are anticipated to assist and perform over 60 percent of surgeries. Improvements in motor accuracy are expected to lower complication rates by 25 percent, saving hundreds of thousands of lives annually. Robots in eldercare are projected to fill over ten million roles by 2040, providing companionship, assistance, and health monitoring for aging populations, particularly in regions like Japan, where 35 percent of the population will be over sixty-five by mid-century.

And to feed that aging population, robots capable of precision farming—planting, harvesting, and monitoring crops—are expected to boost yields by up to 70 percent while cutting resource consumption by 50 percent. By 2040, these machines could account for 60 percent of all labor in the global ag sector. This efficiency will be essential as the world's population is projected to reach 9.7 billion, requiring a 50 percent increase in food production to meet demand.

Then there's security, education, transportation, construction, and, of course, space exploration. Peter thinks the first Mars mission will land around 2030, but instead of carrying humans to the new frontier, the first boots on the Red Planet will belong to a humanoid robot. They'll set up habitats, power stations, maybe even hot tubs, making it safe and comfortable for the soft, squishy human settlers soon to arrive.

For certain, Helen Greiner's vision of practical robots has become reality. Just like ReGen Valley turned bioprinting into an assembly line for life, the iRobot Mafia turned robotics into an invasion that spread from home to warehouse to hospital to farm. The memes mutated. The machines multiplied.

Now, the question is how will we evolve? Every burst in speciation brings with it extinction. The new wipes out the old, and studies show that 47 percent of US jobs could be automated by 2030, which sure sounds like a lot of extinction. The opportunity is hidden in the challenge. If robots are taking on the dull, dirty, and dangerous, what remains are the creative frontiers—the work that draws on what is most distinctly human.

Ben Lamm and the Adjacent Possible

The 47 percent of US jobs that could be automated by 2030 sounds like a terrifying prediction. Yet, historically, automation doesn't eliminate jobs; it reshapes them and often into careers that sound like science fiction. Consider this reframe: "By 2030, robots may take your job unloading trucks, cleaning offices, or stocking shelves, but then you'll be free to pursue your childhood dream of becoming an 'artificial womb engineer.'"

In fact, fast-forward to 2030, and let's scroll through a job board:

De-Extinction Specialist

Location: Global Biodiversity Lab, Iceland
Description: Lead groundbreaking efforts to revive extinct species using advanced tools like CRISPR and somatic nuclear transfer. Responsibilities include genome editing, reintroduction planning, and ecological monitoring. Must have experience in synthetic biology and a passion for rewriting the story of life.

Species Restoration Specialist

Location: Arctic Rewilding Center, Siberia
Description: Seeking conservationists to assist in the reintroduction of woolly mammoths to the arctic tundra. Responsibilities include habitat monitoring, herd management, and data collection. Must enjoy cold weather and working with species that haven't roamed Earth for four thousand years.

Mammoth Reintroduction Publicist

Location: Siberia
Description: Help rebrand the Ice Age. Duties include crafting compelling narratives, coordinating press visits to the Arctic, and handling tough questions from ethicists and climate scientists. Experience in PR and journalism preferred. Thick skin and cold-weather gear required.

Artificial Womb Engineer

Location: Global Biotech Solutions, Singapore
Description: Develop next-gen ex-utero gestation platforms to support the growth of embryos from de-extinct species. Experience in bioengineering required. Experience at a petting zoo a plus.

Genetic Rescue Technician

Location: Conservation Genetics Institute, Kenya
Description: Help preserve endangered and de-extinct species by refining genetic diversity. Use gene-editing tools to address hereditary bottlenecks and prevent extinction recurrence. God complex not required.

The point here is that technology always creates more jobs than it destroys. The bigger point is that each role—like "de-extinction specialist"—is a portal into the adjacent possible that produces opportunities beyond the horizon of our imagination.

To see this in real life, meet a real-world de-extinction specialist: Ben Lamm, CEO of Colossal Biosciences, a company with $225 million in funding, a $10 billion valuation, and a to-do list that includes reviving the woolly mammoth, dodo, and Tasmanian tiger.

Lamm didn't invent the concept of de-extinction, but he is the first person to try it at scale. The quest for species revival began in the 1930s, when early cloning experiments made resurrection a speculative possibility. The 1950s discovery of DNA added credibility, but it was Steven Spielberg's 1993 cinematic adaptation of Michael Crichton's book *Jurassic Park* that truly captured our imagination.

Like *Star Wars* before it, *Jurassic Park* grounded speculative fiction in contemporary science. Three years later, with the cloning of Dolly the sheep, and six years after that, when researchers used DNA from preserved tissue to revive the Pyrenean ibex—which survived only minutes due to severe lung defects—fact began closing in on fiction.

By the mid-2000s, the *Jurassic Park* fantasy of extracting genomic information from mosquitoes trapped in amber proved impossible (amber is too porous to preserve DNA), yet advancements in genetic sequencing opened the door to species revival. By 2020, CRISPR-Cas9 had revolutionized gene editing, giving us the ability to reconstruct the genomes of ancient animals. That's when Harvard geneticist George Church, who many credit with pioneering the field of synthetic biology, proposed rewilding the arctic tundra with de-extincted woolly mammoth hybrids—the region's original keystone species—to restore grasslands, preserve permafrost, and slow climate change. All Church's proposal needed was an entrepreneur willing to attempt resurrection at scale.

Enter Ben Lamm.

A serial entrepreneur with successful exits in multiple domains, Lamm heard about Church's work in genomics and called him up to discuss the intersection of software and synthetic biology, believing this convergence hid the next tech revolution. "That conversation lasted about ten minutes," explains Lamm. "Then I asked George what else he was working on. He [told] me about all the incredible things his labs were doing—everything from neural regeneration to food security to combating climate change. He ended the call with, 'Oh, and I'm working to de-extinct woolly mammoths to preserve the arctic tundra and maybe help elephants and other species.'"

Lamm was thunderstruck by the idea. He stayed up all night researching the possibility of resurrecting the dead. What would happen, he started to wonder, if he applied Spielberg's business model—using vanished species as a tourist attraction—to the problem of de-extinction?

The answer: You can raise a lot of money in a hurry.

The influx of cash allowed Lamm and Church to build on the work of Crichton and Spielberg, expanding the use-case for existing technologies into Hollywood-esque realms. CRISPR, originally developed to treat genetic disorders, was adapted to edit degraded DNA. Tools for sequencing modern genomes were retooled to reconstruct ancient genetic blueprints. Drones and biosensors, staples of agriculture and conservation, were upgraded to monitor rewilded species, collecting data on behavior, health, and ecological impact. Finally, artificial wombs created for neonatal care were refined for species revival—and that includes the twenty-two-month gestation period for a woolly mammoth.

In short, all the jobs we imagined earlier are the adjacent possible unlocked by the work of Lamm and Church. Each of these frontiers comes with a bevy of new revenue streams—with ecotourism being the *Jurassic Park* example.

But consider that *Jurassic Park* example.

It's been four thousand years since anyone has seen a woolly mammoth. It's been more than three hundred years since the 1662 sighting of the last dodo. Imagine a herd of mammoths under an arctic sky or a flock of dodos waddling across a tropical beach. The data collected on biodiversity, ecosystem resilience, and climate adaptation alone would be worth a fortune. And the revenue streams for rural communities created by de-extinction preserves could spread that abundance around.

What began as a conversation between a geneticist and an entrepreneur has become a de-extinction engine, powering novel industries, producing new jobs, and re-creating lost worlds. Colossal has eight- and nine-figure deals with two different countries with red-listed (in danger of extinction) keystone species (critical to the ecosystem) of crucial cultural importance (like, say, the bald eagle in the United States). Yet despite the enormous price tags for these projects, both are cheaper and faster than existing conservation efforts. "In one case," Lamm explains in an interview with the authors, "we're cutting down a country's work by twenty years and saving them hundreds of millions of dollars."

In April 2025, Lamm announced the first de-extinction miracle: the return to Earth of three dire wolf pups after a twelve-thousand-year hiatus. The birth of these pups, named Romulus, Remus, and Khaleesi, was an event celebrated on the cover of *Time*—and for good reason.

Two hundred and fifty thousand years ago, the dire wolf had a home range that stretched from Canada to South America. Extinction occurred ten to thirteen thousand years ago. All that was left behind were fossilized remains.

Lamm's work started with DNA extracted from a thirteen-thousand-year-old tooth and a seventy-two-thousand-year-old skull. Colossal's scientists compared this ancient DNA to the genome of the gray wolf, the dire wolf's closest living relative (99.5 percent genetically identical). They identified twenty differences across fourteen genes that account for the dire wolf's distinctive characteristics—its greater size, white coat, wider head, larger teeth, more powerful shoulders, and more muscular legs.

Rather than a biopsy, Colossal used a blood draw to harvest endothelial progenitor cells—which are cells that repair blood vessels—from living gray wolves. Next, they edited fourteen genes to express their twenty desired dire wolf traits. The dire wolf has three genes that code for coat color, for example. In gray wolves, these genes can cause deafness and blindness. So Colossal engineered two new genes to deactivate these old genes, getting the right coat color without any harmful side effects.

Finally, the edited nuclei were inserted into denucleated gray wolf eggs. Forty-five embryos were transferred into two domestic dog surrogates, with one embryo taking hold in each. After sixty-five days of gestation, Romulus and Remus were born via C-section. Months later, a third surrogate produced Khaleesi.

Once again, dire wolves roam the earth—which is both a statement of fact and a fact that's hard to wrap your head around. Yet Lamm's vision for Colossal's future is even more colossal. "We're down to twenty thousand blue whales or so," he explains. "Imagine stadium-size grow tanks for ex-utero development that allow us to gestate a thousand blue whales at once." Colossal also plans to open-source many of the foundational de-extinction technologies to other conservation groups, so perhaps they won't be the only ones building stadium-size ex-utero gestation facilities.

As far as de-extinction timelines go, Lamm believes we'll see the first woolly mammoth by 2028, the dodo could arrive as soon as 2026, and the Tasmanian tiger even sooner. The real-life Jurassic Park? According to Lamm, unless someone finds a fully frozen dinosaur somewhere—you can't extract DNA from fossils—that idea remains in the "too crazy for now" category. But, as Peter likes to say: "The day before something is truly a breakthrough, it's a crazy idea."

A Moment of Not Zen, Redux

Yo, miracle worker.

Check this out: You coexist with the dead. De-extinct species now amble across eco-preserves near you. And if a Tasmanian tiger happens to eat your liver, Dean Kamen can grow you another, and the robo-surgeon offspring of the iRobot Mafia can stitch it into place.

This is going to be daily life in the middle twenty-first century. Before too long, "I got to pet a woolly mammoth" is going to be a thing that happens over summer vacation. Replacement body parts will become standard fare as well, arguably covered by health insurance. And who knows? Maybe robo-surgeons will start making house calls.

We're normalizing the miraculous. Flashy rock stars won't have pet tigers anymore. Pet saber-toothed tigers? Coming soon.

Planet GPT

So far, this chapter has tracked what happens when AI collides with other exponentials, turning science fiction into science fact at speed and scale. Organ regeneration. Robot speciation. Species revival. Miracles all—and now mass producible.

Scale is the real force multiplier for transforming breakthroughs into abundance. And no technology has scaled faster, or wider, than sensors and networks. What began as the "Internet of Things" (IoT) is fast becoming the "Internet of Everything." Or, as satellite entrepreneur Will Marshall likes to call it, Planet GPT.

This story starts in the early 2000s.

Will Marshall was a NASA astrophysicist with an annoying colleague who kept waving his phone around and griping: "This thing has everything a satellite has—radios, cameras, GPS, rate gyros, accelerometers, processors—and it costs $500 to $1,000." Meanwhile, NASA was building satellites that cost $500 million to $1 billion. "So," explained Marshall on Peter's *Moonshots* podcast, "the question became: 'What are those other six zeroes doing for us?'"

The comment shifted Marshall's thinking and, as a result, the entire trajectory of the space industry. "NASA was interested in inventing everything for itself," Marshall continues. "It was the culture. But top R&D dollars weren't at NASA anymore—they were at Google and Samsung. Miniaturization and increased performance changed the game."

Marshall and his colleagues decided to see if you could really use a cell phone as a satellite. First, they placed a phone in a vacuum chamber. No problem. Next, they strapped it to the side of a rocket. It survived. Finally, they smuggled three phones onto a NASA satellite launch. "NASA freaked," Marshall admits. "I nearly got fired. But it worked. We proved we could take pictures from space for a few thousand dollars."

Inspired by the results, Marshall, alongside Robbie Schingler and Chris Boshuizen, left NASA and co-founded Planet, a company transforming Earth observation with smartphone-inspired microsatellites called CubeSats. Affectionately, they're known as *doves*.

Today, Planet has some five hundred doves in orbit, with two hundred devoted solely to planetary imaging. Each dove is tiny, weighing ten to fifteen pounds at most, yet capable of transmitting two gigabits of data every second. The majority provide a three-by-three-meter view of Earth. Twenty are capable of fifty pixels per centimeter—or what it takes to count stalks of corn in a field.

The fleet represents a seismic shift for the space industry: a fourfold reduction in rocket launch costs, a thousandfold increase in satellite performance, and a ten-thousand-fold increase in data-transmitted-per-dollar. That last metric is the real game-changer, creating "a space renaissance," according to Marshall, and transforming satellite technology into an engine for global insight and action.

So, what does action look like?

Planet believes transparency equals accountability. Marshall feels there are three areas where his technology can make the biggest difference: digitalization, sustainability, and peace.

Space-based digitalization measures things we've never been able to measure before. Agriculture is Marshall's go-to example. "We image every farmer's field on Earth every day. That's roughly one-quarter of the planet. We can tell what crop is growing, how well it's doing, what practices the farmer is using, whether they should add water, is the crop in need of fertilizer, or herbicide, or not. In total, we think we can increase crop yields by 20 percent and decrease herbicides, fertilizers, and other inputs by 20 percent." When you get a 40 percent improvement in a trillion-dollar sector? As Marshall says, "That's a big deal."

On the sustainability front, transparency has already led to accountability. Planet does weekly scans of the Amazon to detect new road starts—often indicators of illegal deforestation, mining, or narcotics operations. "In 2023," he explains, "our data supported three thousand missions by the Brazilian government, contributed to a 55 percent reduction in deforestation [according to the UN], leading to nearly $2 billion in asset confiscations."

Planet also maps carbon stocks and biodiversity hot spots, which is creating new markets for ecosystem services via carbon credits and biodiversity offsets. "Half the world's GDP depends on nature," says Marshall. "We've wiped out 70 percent of the planet's biodiversity in just forty years. If we want an economy that lasts, we need to rebuild that foundation—and you can't manage what you can't measure."

On the peace front, Planet regularly shares war zone images with major newspapers, so claims by combatants about "not bombing civilian targets" can be independently validated. Their satellites also track troop movements, detect early signs of environmental disasters, and monitor refugee camps. In 2024, when Germany experienced historic flooding, Planet directed first responders, coordinated relief efforts, and assessed damages. "We're building an [imaging] system to manage Earth's resources," says Marshall, "much like the eye and visual cortex help humans make smart decisions."

Looking ahead, optical improvements will extend satellite imaging into other spectral bands. Minute changes in soil moisture, early threats to vegetation health, microscopic shifts in atmospheric composition—soon, all will be measurable. The result is Planet GPT, a dynamic dashboard for managing Earth's ecosystems with precision and transparency. "This [shift] is a little like the mainframe computer to the desktop computer revolution, but for space," explains Marshall. "And like the computer revolution, the upshot of the space revolution is a data revolution."

Planet is democratizing accountability. It's providing a new level of information abundance, ushering us toward an era when you can know anything, anywhere, anytime. Want to stop illegal poaching? Want to know the largest polluter in your hometown? Or even what the kids are wearing on the streets of Singapore? Information abundance means empowerment. For a long time, only governments and wealthy corporations had access to this data. Today, these insights belong to all of us.

Omniscience—it's what's for breakfast.

Powering Africa

Energy abundance is the true turning point for the world. Electricity is the über-enabling technology. If you light up a village, you ignite education, healthcare, commerce, community—this list goes on.

Nowhere does this matter more than in Africa.

When we published *Abundance* in 2012, Africa's grid was a patchwork of inadequacy. Six hundred million people lacked electricity. That's 60 percent of the continent. In rural areas, the number climbed above 80 percent. Even the solutions we offered felt like fables—solar-powered villages, smart grids in the savanna, affordable energy for all.

Yet Africa receives over sixty million terawatt-hours of solar energy annually, the equivalent of 7.6 million nuclear power plants. This is both more than any other continent and a sum that vastly exceeds the world's energy needs. In some regions, daily solar radiation reaches five to seven kilowatt-hours per square meter, which is enough to power entire economies. With sunlight, like so many resources, it's not a question of scarcity, it's about accessibility.

Take Gbamu Gbamu—pronounced *Bam Bam*—a rural community in southwest Nigeria. For centuries, the lights never went out in Gbamu Gbamu because they were never on. The 3,500 residents lived in darkness, the night punctured only by the glow of kerosene lamps and the roar of generators.

Everything changed in 2018.

An eighty-five kilowatt mini-grid installed by Rubitec Solar turned Gbamu Gbamu into an oasis of electricity. Farmers began to irrigate with solar-powered pumps. Children could study at night. Shopkeepers sold cold drinks from new freezers. The health clinic refrigerated vaccines and ran medical equipment without interruption. And this metamorphosis didn't rely on Nigeria's notoriously unstable power grid—which loses a World Bank–estimated $29 billion annually to blackouts and intermittency. Instead, it came from a compact, decentralized system humming quietly in the darkness on the edge of town.

The story of Gbamu Gbamu is a testament to energy transformation and the power of mini-grids. Yet for much of sub-Saharan Africa, similar stories are rare exceptions. Energy poverty traps millions in its cycle of hardship, stealing time, resources, and potential. Each year, six hundred thousand lives are lost to indoor air pollution, and billions of dollars are wasted on unreliable energy sources.

Mini-grids are a potent solution. According to the World Bank, they're the most cost-effective way to electrify remote communities. Today, they power the lives of forty-seven million people. By 2030, they're projected to serve half a billion. As solar mini-grids also reduce emissions, the International Energy Agency estimates that they can cut 1.5 gigatons of carbon by 2030. It's the equivalent of removing over three hundred million cars from the road.

This transformation is already underway. A woman named Damilola Ogunbiyi sits at the center of the story. She's the managing director of Nigeria's Rural Electrification Agency, overseeing a $550 million attempt to turn on the lights. Because of her work, more than five million people in rural Nigeria—including the residents of Gbamu Gbamu—access clean electricity via solar mini-grids and stand-alone smart systems. These systems use IoT sensors and digital meters to track electricity production, energy consumption, and maintenance needs.

More importantly, these systems also enable mobile payments, one of the bigger weapons in the fight against energy poverty. Pioneered by companies like M-KOPA in Kenya and Fenix International in Uganda, these pay-as-you-go solar systems give low-income communities access to power without huge up-front costs. Nigeria followed suit. Families now pay for electricity in increments that fit their budgets—and the lights stay on.

In Gbamu Gbamu and thousands of other communities, microgrid systems provide power and create opportunity. The International Renewable Energy Agency found that doubling the global share of renewable energy by 2030 would increase annual global GDP by $1.3 trillion. In sub-Saharan Africa, access to modern energy services already boosts GDP by 2–3 percent. And electrified villages see improved health outcomes, better educational opportunities, and a sharp rise in the number of small businesses.

This is also why Ogunbiyi's work hasn't stopped at Nigeria's borders. As CEO of Sustainable Energy for All and the UN secretary-general's special representative for sustainable energy, she locked up $1.3 trillion in finance commitments to support renewable energy in over ninety countries. She's building ecosystems for energy equity. By 2030, solar mini-grids could bring affordable electricity to 265 million Africans and meet nearly half the continent's energy needs.

Yet no single narrative captures this entire picture. Yes, Damilola Ogunbiyi is lifting hundreds of millions of people out of energy poverty. Yes, Dean Kamen is bringing organ regeneration to the masses. Yes, Ben Lamm and George Church are de-extincting woolly mammoths to fight climate change and combat biodiversity loss. But these aren't isolated incidents. They're exemplars of larger trends.

Ogunbiyi's solar revolution, for example, is one thread in a much broader fabric. Since 2011, global solar capacity has surged over 700 percent. Costs are down 90 percent and still falling. Lithium-ion battery prices, once the Achilles' heel of renewable energy, plummeted 97 percent over the past few decades. Blend these developments with advances in AI-energy management and mobile payments and you have the recipe that brought solar to scale. The result? One person can now light up a continent. One lab can now regrow body parts. One company can now resurrect the dead. The miracles are mounting.

But when anyone and everyone has access to godlike tools, what happens when things go wrong? "When you invent the ship, you invent the shipwreck," as French philosopher Paul Virilio pointed out. Lately, the shipwreck has been making headlines. AI arms races. Rogue actors. Climate chaos. Techno-unemployment. These sound like more examples of our growing obsession with the apocalypse. They're anything but—and that's exactly the problem.

Welcome to the dark side of abundance.

CHAPTER SIX

The Dark Side of Abundance

So far, we've told tales of triumph. It's been an ascent of mythological proportions. Exponential tech. Heroic entrepreneurs. A world transformed. Every mind on Earth now connected. Every corner of the globe now visible. We solved for famine, cured deadly diseases, and resurrected the dead. But paradise comes at a cost.

This chapter is about the price tag.

The Price of Paradise

Let's take a walk through New York City circa 1894. The streets are rivers of horse manure: three million pounds produced daily by one-hundred-and-fifty-thousand horses. It's a quagmire of crap. The air is thick with coal smoke. Life expectancy hovers around fifty years. If you're lucky enough to have a job, you're likely working eighty-hour weeks. If you get sick, your best medical option is probably bloodletting.

Welcome to the good old days.

Of course, the manure problem wasn't just a byproduct of city life—it was a side effect of abundance. Four thousand years earlier, humans domesticated the horse. That single breakthrough unlocked faster trade, faster travel, and the modern city. But combine horse-based transport with the scale of a city like New York, and you get a new kind of problem: an abundance of crap, literally piling up in the streets.

It took millennia for this abundance problem to emerge. But once it did, we solved it in a decade with the internal combustion engine. Nationally, the horse population peaked at 26.5 million in 1915, but numbers plummeted as cars grew from 8,000 in 1900 to 26.3 million by 1930. In major cities, the horse-to-car ratio flipped within a decade, with a 90 percent decline in urban working horses by the early 1920s. It was a miracle solution that, in hindsight, was also the trigger for a much larger problem: the climate crisis.

This is the pattern. It's the abundance paradox all over again. Each new solution unleashes a host of new problem. Only now, it's accelerating. In the twenty-first century, the issues of abundance aren't arriving in four thousand years or even forty. Try forty days. That's about how long it took for social media to impact mental health. For AI to fake voices. For large language models to flood the web with synthetic noise.

We've flipped the script.

In *Abundance*, we focused on the upside of exponentials and the promise of "more." That promise still holds. But so does the paradox: The faster we solve old problems, the faster new challenges emerge, ones we don't fully understand and may not yet be prepared to handle.

Yet, the same tools that create these dilemmas may also give us the ability to solve them. Would you rather face today's challenges with today's tools, or today's challenges with 1900s' tools? When we ask this question to audiences, the answers always favor the present tense and the power of exponentials.

As we turn our attention to the dark side of abundance—runaway AI, bioterror, fragile democracies—it's helpful to remember this pattern and its potential. This also explains our aim. We need to explore these problems. We also need to remember the same forces that create them may also hold their solutions.

Welcome to paradise, where the future isn't just better than you think. It's more complicated than you might imagine.

The Problem of More

Abundance isn't free. For all the breakthroughs we've explored, there's a growing class of problems. These are new categories of unintended consequences born from exponential acceleration.

In the appendix, you'll find two sets of charts. The first is the evidence for abundance. It tracks our escape from scarcity. The second reflects the dark side of abundance—call it the fine print of the law of more. Or, as biologist E. O. Wilson once said, "The real problem of humanity is the following: We have Paleolithic emotions, medieval institutions, and godlike technology."

Here's our current top ten list of optimism's biggest adversaries and the unintended consequences of "more":

1. Climate Crisis
2. Obesity Epidemic
3. Drug Abuse and Overdoses
4. Mental Health Crisis and Information Overwhelm
5. Microplastics and Forever Chemicals
6. Human Population Flux
7. The Growing Rich-Poor Divide and Political Polarization
8. Biodiversity Loss and Ecosystem Collapse
9. The Threat of Future Pandemics and Bioterrorism
10. Dangers of AI

If your favorite disaster du jour is not on this list, please forgive us. Yet, as we'll see, if you're looking for grounds for data-driven pessimism, this top ten list gives you plenty to consider.

The Heat Is On

Greenland's Jakobshavn Glacier was once a fortress of time, its ice solidified into place over a hundred thousand years ago. Each layer is a page from the Earth's history: trapped water, ancient pollen, volcanic ash sealed in crystal. The glacier should stand as a monument to the old ways: stability, silence, and slow change.

Not anymore. Now the ice is on the run. Soon, it may be nothing more than a memory.

Today, the Jakobshavn Glacier retreats a hundred and thirty feet a day. In the past two decades, this single meltdown has dumped enough water into the ocean to fill Lake Erie—twice.

Scientists call Jakobshavn "the canary in the coal mine of climate change." Glaciologist Jason Box calls it "a sleeping giant waking up." Because, of course, it's more than frozen water that is vanishing.

The loss of ancient ice disrupts sea levels, ocean currents, marine life, and climate feedback loops. It's a crushing blow to arctic ecosystems. If this meltdown continues, there's a real risk we'll tip ourselves past the tipping point—beyond which the damage is irreversible.

Welcome to the bad new days.

Once an academic prediction, the climate crisis has become a global transformation. Since the Industrial Revolution kicked off our love affair with fossil fuels, we've nudged the global thermostat up 1.1 degrees. The atmosphere now contains more carbon dioxide than it has in eight hundred thousand years. We're conducting a chemistry experiment with our planet. Literally.

Perhaps another chemistry experiment can save us. Consider the $100 million Carbon Removal XPRIZE, which attracted over twelve hundred teams from thirty countries—the largest response in the XPRIZE's twenty-eight-year history. And these aren't garage tinkerers. They're leading entrepreneurs wielding advanced technologies that pull carbon from the ocean and atmosphere. Based on historical metrics, this single competition will spawn $20 billion worth of climate-focused companies.

There's also tangible progress. In March 2025, the first $50 million of the prize was awarded to Mati Carbon for "enhanced rock weathering," which involves spreading crushed basalt over agricultural lands to significantly increase carbon absorption while restoring soil health, water retention, and yield size. In short, it lets farmers fight climate change while earning more from their fields.

Yet carbon removal is one piece in a larger environmental puzzle. Another is coping with extreme weather events. Take wildfires. In the past few decades, a natural ecological process morphed into a succession of disasters. In 2023, wildfire smoke turned New York into something out of *Blade Runner*, while Greece watched flames devour ancient forests. In 2025, a large chunk of Los Angeles was reduced to ash. The price tag? In the United States alone, wildfires produce $350 billion in damages every year.

Yet here, too, there are emerging solutions. Consider five seemingly unrelated fire issues: the need to replant trees after a burn, the need to protect structures during a fire, the need to restore soil health to the forest, the need to build firebreaks during a fire, and finally the need to haul large quantities of woody biomass out of rugged and roadless mountains.

Now consider drones. Already, tree-planting drones aid in post-fire reforestation. The Reno-based start-up Flying Forests has a drone that fires "seed-bullets" into the ground and plants 30,000 trees a day, compared to 400–1,200 for hand-planting. With a little innovation, these same drones can be built modularly, with snap-on parts. In spring, we have a tree-planting drone. In summer, during fire season, this same drone can be augmented with heat-shielding and a modified payload delivery system to launch flame-retardant missiles. In the fall, by slightly altering the payload again, it can be used for soil restoration. Finally, there's woody biomass removal. While the lifting capacity of a single drone is under ten kilograms, a swarm of drones positioned around a haul-bag can do the job. This single solution—a drone—solves five intractable fire and forest health problems.

A second solution is the $11 million Wildfire XPRIZE. The way to win the prize? Develop an autonomous system that can monitor a thousand square kilometers and detect and extinguish fires within ten minutes of ignition. Imagine a network of satellite-based, AI-powered sentinels watching over our forests and towns, ready to pounce on any flame larger than three meters in diameter.

These developments are fuel for optimism. The convergence of renewable energy, electric vehicles, materials science, robotics, and AI is birthing a trillion-dollar industry of climate solutions. Bill Gates predicts that the scale of innovation is massive: "There will be the equivalent of Microsoft, Google, Amazon-type companies that come out of this space."

But is it happening fast enough?

If temperatures rise another 0.9 degrees and hit two degrees Celsius above preindustrial levels, we're looking at the loss of 99 percent of our coral reefs, the collapse of entire ecosystems, and a sea-level rise that could displace nations. By 2050, hundreds of millions of people will be pushed into climate-related poverty.

It's a race against time. Mostly, it's a race against ourselves. The technology exists to create solutions. But solutions aren't enough. We also need cooperation. Removing the excess carbon in our atmosphere requires more than a string of individual efforts. We need collaboration on a global scale, and we need it now.

The Plasticene Era

Plastic—it's the miracle material of the modern world. Cheap and durable. The Tupperware of Tomorrowland. It keeps things fresh and clean. It's light. It's strong. And now it's inside you.

In 2022, Dutch researchers discovered microplastics in human blood samples. Tiny fragments smaller than a red blood cell showed up in seventeen out of twenty-two healthy adults. A few months later, we found microplastics in human breast milk. Then in placentas. Then in lung tissue. In 2024, researchers discovered there are seven grams trapped inside each of our brains—the equivalent of a plastic spoon.

Welcome to the Plasticene Era, where an abundance of convenience is killing the planet. We've shrink-wrapped our world. Every minute, a garbage truck's worth of synthetics is dumped into our waterways. By

2050, the weight of all the plastic floating in the oceans will be greater than the weight of all the fish.

And this is the crisis we can see. PFAS are the invisible invasion. PFAS, or per- and polyfluoroalkyl substances, are forever chemicals. They don't break down, not in the environment, not in our bodies. First developed by 3M and DuPont in the 1940s, today these compounds are everywhere: the nonstick coating of your frying pan, the stain-resistant treatment on your carpet, the waterproof shell of your rain jacket. They're in firefighting foam, grocery store packaging, and dental floss. Now, they've infiltrated our food chain and our bodies.

According to the CDC, you can find PFAS in the blood of 97 percent of all Americans. It's estimated that we consume about a credit card's worth every week through our food, our water, and the air we breathe. They've been detected at the summit of Mount Everest, in the depths of the Marianas Trench, and, according to a 2023 study, in rainwater on every continent, including Antarctica. It's official. We've managed to contaminate every corner of the planet with a class of chemicals designed to never break down.

PFAS are killers. They've been linked to cancer, liver damage, hormone disruption, decreased fertility, and increased obesity. They blunt the immune system, reduce the effectiveness of vaccines, and may contribute to attention deficit disorder. Over six hundred species are suffering. PFAS cause lesions in alligators, thyroid problems in seals, and reproductive problems in dolphins. Once again, the unintended consequences of abundance have brought us to the brink.

Yet here, too, there's hope. Governments are beginning to act. The European Union has already banned thousands of PFAS, and similar legislation is gaining traction in the United States. Major companies started to phase out production in 2025. Sensors and networks are also in play. The Helmholtz Centre for Environmental Research has an AI that tracks microplastics using satellite data, creating the first map of plastic waste distribution and helping direct cleanup efforts and identify major sources of pollution. At UCLA, engineers created a "super-sponge" that removes 99 percent of PFAS from water in seconds, and can be wrung out and reused hundreds of times. And scientists at the Wyss Institute engineered enzymes to break down PET plastic—one of the most common forms of plastic pollution—in hours instead of centuries. "We're essentially teaching biology to eat our mistakes," explains Dr. Ting Xu, who led the research team.

The question now is whether our solutions can scale faster than our pollution. We produce about four hundred million tons of plastic annually, and that number is expected to double by 2050. If we want to stay ahead of the crisis, the best hope is to move away from the material. Here, biodegradables and circular economy solutions are creating alternatives to traditional plastics that go beyond recycling. Think enzyme-based plastics designed to break down into chemicals that are good for the planet, and closed-loop manufacturing systems where yesterday's debris becomes tomorrow's raw material. Once again, the challenge is speed. Can innovation outpace devastation? Can global cooperation prevent worldwide collapse? We like our chances. But no question about it—the race is on.

A Paradox of Plenty

Our ancestors battled starvation. We battle the refrigerator—and starvation. It's a pair of contradictory crises. Nearly a billion people are hungry. Another 2.6 billion are overweight or obese. We live in a world where feast and famine coexist on a colossal scale.

The story starts on the African savanna, where our predecessors played a deadly game of nutritional chance. Some days brought bounty, others brought nothing. The people who survived became expert energy misers, storing today's calories against tomorrow's famines and the outcome of what geneticist James Neel called the "thrifty gene hypothesis." Those who hoarded best got to pass their genes on to the next generation. It was a great way to survive in prehistory. It's a terrible strategy in the age of Pop-Tarts.

Sugar, for instance, was a rare commodity. For most of history, the sweetest thing anyone tasted was a piece of fruit. Then came European colonial expansion and the first large-scale sugar plantations in the Americas. The real rush hit in the twentieth century, when food processing became an industrial art form. Sugar started sweetening everything from our morning coffee to our evening cocktails, and we never stood a chance. Our bodies are playing by Stone Age rules in a sugarcoated, space-age world.

Today, 38 percent of the world's population is overweight or obese. By 2035, the figure is projected to rise above the 50 percent mark. It's a health crisis and an economic time bomb. The global cost of obesity was $1.96 trillion in 2019. By 2035, that number is projected to hit $4.32 trillion—or roughly 3 percent of global GDP.

The challenge of obesity is abundance gone awry. The solutions require a multitiered approach: medical interventions to reprogram our genetic software, educational initiatives to rebuild nutritional literacy, and policy reforms to reshape our food landscape.

On the medical front, GLP-1 agonists are the latest reprogramming tool, with studies showing that people taking these drugs can drop 10–15 percent of their body weight in a year. At the same time, start-ups are tackling the problem with an exponential arsenal ranging from gut microbiome manipulation to AI-powered personalized nutrition. It's not the first time we've tried to fight our weight with technology, but it may be the most successful.

In education, we're making less progress. Nutrition literacy remains dangerously low. Policy interventions have helped. Sugar taxes in Mexico and the UK led to a steady decline in sweetened beverage consumption. The same is true for front-of-package labeling, junk food advertising restrictions, and subsidies for healthier foods. But today's pilot programs need to become tomorrow's cultural revolutions, because the longer we delay, the heavier the consequences.

Yet consider the flip side: While billions struggle with their weight, hundreds of millions fight starvation. According to the UN, 735 million people face hunger every year, with nearly half living in "acute food insecurity." Against this backdrop, global food waste now exceeds 1.3 billion tons a year, which is enough to feed every hungry person on the planet three times over.

It's almost a tale of two worlds. It's definitely another abundance paradox. Since 1961, while the world's population doubled, cereal production tripled and meat production more than quadrupled. It's one reason why, since the late 1990s, Americans spend less than 11 percent of their disposable income on food—down from 25 percent in 1929. The problem is fragile regions with vulnerable economies. Families in Yemen, for example, can spend nearly 70 percent of their income just to avoid starvation. As science fiction writer William Gibson famously said, "The future is here, it's just not evenly distributed."

Distribution is coming. It's being called a "Second Green Revolution." It's really the convergence of exponential technologies that are reinventing how we grow, distribute, and define food. Already, we explored revolutions in cultured meat, AI-driven precision farming, and PR23 hybrid rice. But this is only a peek at the pipeline.

Vertical farming companies like AeroFarms and Plenty move farming from the country to the city. We now grow crops in warehouses, using 95 percent less water and producing yields 390 times higher per square foot. In 2023, the industry hit $4.8 billion. By 2030, it's projected above $33.5 billion.

Israeli start-up Beewise deploys robotic beehives to combat colony collapse disorder. Companies like Ginkgo Bioworks program microbes to create novel proteins and flavors, essentially teaching biology to cook. Organizations like AgriLedger use blockchain to add transparency to our food supply, increasing farmer incomes by up to 300 percent in pilot programs in Africa.

Is this enough?

Open question. Political instability continues to disrupt distribution networks. Cultural barriers slow the adoption rates for novel food technologies. And by 2050, climate change will threaten up to 30 percent of global crop yields.

The answer may lie in what food scientist Dr. Bruce Friedrich calls *the protein portfolio approach*: a strategy that treats protein the way finance treats risk. Instead of betting on a single food source, we diversify, combining animal proteins, plant proteins, and next-generation alternatives like fermentation-derived and cultivated proteins, each optimized for different environments, cultures, and constraints.

The result is a food system that is harder to break. When climate shocks wipe out crops, fermentation keeps running. When supply chains fracture, local protein production fills the gaps. When culture resists one solution, another fits. By using every tool in our exponential arsenal, the protein portfolio approach doesn't just increase abundance, it builds resilience into the system itself.

The Age of Engineered Addiction

The age of abundance is an age of addiction. It didn't happen all at once. It didn't happen by accident. It happened because everything designed to hook human attention—screens, social media, streaming, gaming, designer drugs—hit an exponential curve at the same time.

Over the past few decades, our compulsive behaviors have been tracked, tweaked, and scaled by artificial intelligence. We're pumping out synthetic highs chemically engineered for maximum potency and minimum cost. Algorithms. Labs. Market forces. All of them converged on the same target: the human brain. It's a full-force rewiring of pleasure on a planetary scale.

Social media was the first wave. Billed as a connection tool, it became an attention predator. Platforms built to facilitate communication became tools to supercharge addiction. Insta-Google-X-Face overrode occasional boredom with ceaseless content. Digital validation replaced real connection. Then along came AI, which tracked us with a single aim: Turn habit into compulsion. Echo-chamber engagement. Amplified outrage. Weaponized comparison. Distraction became dependency—and just like that we were adrift in a sea of mental health disorders.

By 2022, 95 percent of teenagers were on social media, and prey to the addictive nature of digital feedback loops. Kids lost resilience: the ability to tolerate discomfort; sit with boredom; or navigate the slow, messy, and unpredictable nature of human relationships. Between 2012, when *Abundance* hit the stands, and today, the number of sleep-deprived high school students almost doubled; anxiety and depression each jumped by nearly 25 percent; and for those between the ages of ten and twenty-four, suicide became the second leading cause of death.

It was an entire generation unmoored by algorithms, rewired for escape, and Big Pharma to the rescue. In the late 1990s, pharmaceutical companies began marketing opioids as safe, nonaddictive pain relief. OxyContin flooded the market on a wave of doctored research, aggressive lobbying, and social media ad campaigns. Overdose rates reached record highs before regulators cracked down. But the addiction didn't end when the prescriptions ran out. Heroin became the fallback. Then fentanyl, the ultimate designer drug.

Fentanyl didn't require fields of poppies or fleets of smugglers. It was synthetic and strong. A $5,000 kilogram bag could be pressed into half a million pills worth millions. The cartels didn't need street dealers anymore. They upgraded—to Snapchat, Telegram, and Instagram.

And fentanyl delivered. Fifty times more potent than heroin, and often lethal. In 2022, the DEA seized enough fentanyl to kill every American—twice.

Social media and synthetic opioids aren't separate crises. They're two sides of the same system. One primes the brain for addictive behavior, the other exploits it. Ubiquitous screens, social media, attention economies, algorithmic distribution, designer drugs—they all converge. Digital addiction fuels mental health collapse, which fuels substance abuse, which fuels more digital addiction. It's a closed system of engineered compulsion.

If technology created this crisis, it might also help us solve it— by intervening at every level where addiction takes hold: biology, psychology, behavior, and supply chains. There's an anti-fentanyl vaccine developed at the University of Houston, which blocks the drug from crossing the blood-brain barrier, and psychedelic therapies that rewire addiction itself, treating the cause and not the symptom. New AI models also show promise, scanning social media posts, detecting early signs of depression, and flagging at-risk users. And blockchain-backed prescription tracking could cut off the counterfeit supply chains feeding the fentanyl crisis. Yet the race is on. The next decade will determine whether we take control of the machines of addiction or become the first species to lose our minds to our inventions.

Yet Another Moment of Not Zen

Feel dizzy yet? Drunk on doom? Drowning in existential dread? It's a David-versus-Goliath moment—except there's an army of Goliaths.

This is more than an information-processing problem. It's a devastation-processing problem. Every day brings another headline: hotter, lonelier, sicker, louder. Systems unraveling. Breakthroughs reborn as threats. And the truly weird part—you're expected to clean out your inbox. Walk the dog. Smile during meetings. Remember what it felt like to be normal.

The self-help columns have advice: Darkness retreats. Dopamine fasts. Cryotherapy. And when all else fails, Netflix and chill.

It's a Band-Aid on a heart attack.

So no, this is not a moment of Zen.

But it is a call to action.

The Fatal Unraveling

In 2019, the last female Yangtze giant softshell turtle died in a Chinese zoo. She left behind three surviving members of her species. Three. And they are all male. So, when these fellas die, that's the end of the story. Her death wasn't an anomaly. It was a preview. Our planet is hemorrhaging species at a rate not seen since an asteroid wiped out the dinosaurs. This time, the asteroid is us.

Since 1970, wildlife populations have declined by 69 percent. Die-off rates are a thousand times greater than normal. A 2019 United Nations report found a million plant and animal species are now threatened with extinction. By 2050, somewhere between 37 and 50 percent of all life on Earth will be gone.

And scientists aren't just worried about the poster-child extinctions—tigers, elephants, rhinos—they're tracking a deeper unraveling. Pollinators are vanishing. Ocean food chains are fracturing. Microbial ecosystems are collapsing. Stanford biologist Paul Ehrlich says it's like stripping rivets from an airplane mid-flight.

Ehrlich's name may sound familiar. His 1968 book, *The Population Bomb*, was an enormous bestseller. It painted a nightmare of mass starvation and a dying planet crushed by too many humans. But Ehrlich was spectacularly wrong. The population explosion morphed into a slow-motion implosion. Global fertility rates dropped from 5.05 children per woman in 1950 to 2.32 today, and with profound implications.

On occasion, your authors have taken opposite sides of this argument—Steven arguing on behalf of the biosphere, Peter proclaiming population decline an existential risk.

The truth, as usual, is more complex.

The debate between overpopulation and demographic collapse has been raging for years. Some argue we have exceeded the carrying capacity of our planet's biosphere. Others point to the vast swaths of open land available for farming while documenting the worldwide decline in birth rates. Both carry elements of the truth.

The replacement birth rate is 2.1 children per couple. In the ten lowest-birth rate countries, the numbers are far below that line: Taiwan (1.11), South Korea (1.12), and Singapore (1.14) lead the list, with Japan (1.20) and much of Southern Europe following suit. Nations that once feared overpopulation, like China, are scrambling to reverse this trend. Even the United States has felt the impact, with a replacement birth rate of only 1.7 during the decade between 2015 and 2025.

Despite this decline, we're still adding millions of new people to the planet every year. It's a trade-off, as each child born requires resources that other species can no longer afford to lose.

Humans use between 25 and 44 percent of the planet's primary productivity, which is the total energy that plants capture through photosynthesis and the annual "food budget" for life on Earth. In some ecosystems, like temperate forests and grasslands, that number exceeds 50 percent. Why are species die-off rates a thousand times greater than normal? As environmental journalist Richard Manning once wrote: "We [humans] have simply stolen the food, the rich among us more than others."

We've also co-opted 77 percent of Earth's habitable land, with nearly three-quarters of that dedicated to growing the food it takes to feed ourselves. Half of all fresh water is siphoned off for human consumption. And the result is an even more startling number: The total biomass of wild mammals today is just 4 percent of what it was before humans arrived.

We built modern civilization on an expansionist model—more people, more growth, more consumption. But more civilization means less biodiversity, which is a crisis for all life. Forests, wetlands, reefs, and grasslands regulate the climate, purify water, and anchor the food chain. Destroy enough of these ecosystem services and the web of life unravels. Scientists call this a *tipping point*. History calls it *collapse*.

Yet the same forces that drive economic progress also drive population decline. As societies become more resource abundant, birth rates tend to fall. Increased education, especially for women, delays childbearing, while economic opportunity shifts priorities. Historically, large families meant more hands for farming and household labor. Survival was at stake. But in a knowledge-driven economy, children are more of a choice than an imperative.

At the same time, declining birth rates bring their own challenges.

Aging populations strain healthcare systems, social services, and pension plans. In Japan and Italy, small towns are vanishing completely. And the nations that try immigration as a solution face cultural upheavals that challenge national identities and reshape traditions.

In the midst of the upheaval, scientists are scrambling to unravel the unraveling. Already, we examined Colossal Biosciences' de-extinction efforts, but their goal isn't just to resurrect lost species. The aim is to restore lost ecosystems.

And they're not alone.

Geneticists are creating coral reefs adapted to rising ocean temperatures, bioengineering trees that absorb more carbon, and CRISPR-ing disease-resistant endangered animals. Lab-grown meat produced in bioreactors can replace factory farming, just as vertical farming can replace traditional farming, and both without the need for vast amounts of land and water.

Then there's AI. DeepMind's artificial intelligence has mapped over two hundred million proteins, giving us novel insights into how to protect species before they vanish forever. AI is also beginning to model ecosystems. With massive data sets drawn from satellites, sensors, and scientific fieldwork, we can now simulate the cascading effects of extinction, migration, deforestation, and climate change—and perhaps learn to stop the slide.

It's an arsenal of eco-saving technologies, each meant to counteract the destruction we've already caused. But for every species we de-extinct, we are driving thousands more to the edge. Reforestation drones and AI-powered conservation may slow habitat destruction, but until humans return more to the ecosystem than we extract, the best way to halt the unraveling is to reduce our footprint.

For the first time, that might actually be an option. As human numbers decline, humanoid robots are on the rise. Tesla's Optimus, Amazon's Digit, and Figure's Figure 03 mark the first wave of this shift. Each of them presents solutions to the problem of population decline, which also makes them solutions to the problem of biodiversity loss. A 2023 Goldman Sachs report suggests that humanoid robots could fill up to 126 percent of the expected labor shortage in aging nations. In Japan, where nearly 30 percent of the population is over sixty-five, robots already serve as caregivers, security guards, and hotel staff. If automation can offset labor shortages, the need for population expansion fades and we can begin to balance human progress with biosphere preservation.

The Great Leveling

What does *poverty* mean? Once, it meant hunger, a lack of shelter, and a scarcity of material possessions. It was backbreaking labor, no access to education, too much sickness, an early death, and a pauper's grave.

Does this definition still apply?

Certainly, poverty can mean those things, but over the past few decades the struggle for survival has become a moving target. In much of the world, poverty no longer looks like starvation; it looks like obesity. It's not desolation and isolation but an overabundance of social media–fueled connection. It's not a lack of information; it's a deluge of data.

And while we think about inequality in financial terms, this ignores the revolution in human welfare created by technology. For certain, the wealth gap is immense. Today, the poorest half of humanity has access to 2 percent of global wealth, while the richest 10 percent control 76 percent. But what those numbers miss is the great leveling: the most radical expansion of human capability in history.

In 2023, 7.33 billion people, that's 91 percent of humanity, owned a mobile phone. Each device has more computing power than all of NASA during the Apollo mission and accesses more knowledge than was contained in the Library of Alexandria. While traditional measures of inequality show the wealth gap widening, they miss the exponential rise in quality of life that technology provides. Today, affordable technology gives even the poorest access to capabilities that would have been unimaginable for most of human history.

Chart 1
GDP Per Capita

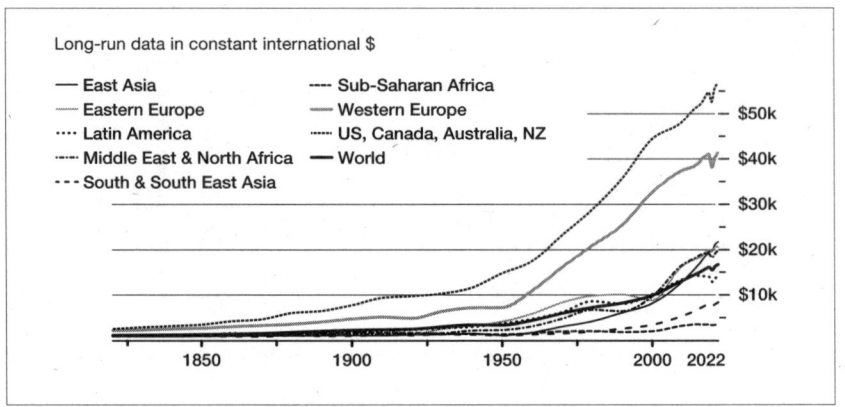

Consider healthcare. If you own a smartphone, you can check your heart rate, track sleep patterns, and get AI-powered medical advice on demand. You can also find any kind of doctor you want via telemedicine portals, have your medicine hand-delivered to your door by a robot, and, by decade's end, be able to run full diagnostic scans from the comfort of your home. Compare that to King Charles II of England, who had the best healthcare available yet still died in 1685 after doctors treated his kidney disease with a cocktail of alcohol and pulverized human skulls.

Education is another example. Anyone with internet access can take courses from Harvard, MIT, and Stanford for free through platforms like Coursera, Khan Academy, and edX. King Henry VIII, despite being one of the wealthiest monarchs in antiquity, had to import scholars from Italy just to learn basic math.

The great leveling doesn't eliminate the problem of inequality, but it does mean that we need a new way to measure progress and prosperity. Consider the figure above, which shows what economists call "the rising floor effect." While the absolute gap between rich and poor may be growing (B > A), the minimum standard of

Chart 2
Rising Floor Effect

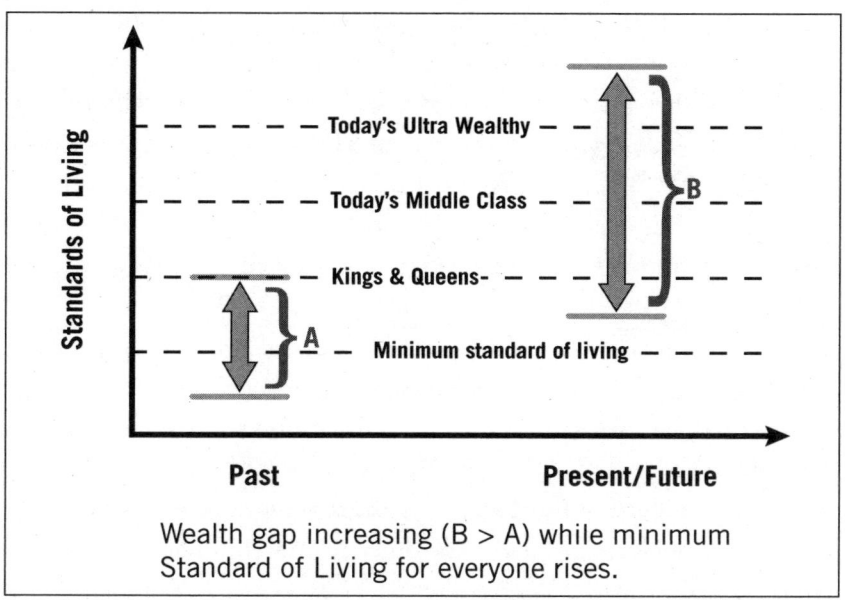

living—the floor—is rising faster than ever, which is why a rural villager today has better access to information, entertainment, and basic services than John D. Rockefeller did at the height of his wealth.

Economic mobility shifts with technology, policy, and the design of our institutions. The wealth gap is a signal that our systems haven't adapted to the exponential age. Traditional paths to prosperity—education, hard work, incremental progress—were built for a linear world. But when wealth accumulates through networks, automation, and financial leverage, these strategies break faster than they can be replaced.

What's the best way forward? The truth is, we don't know. All we know is that the challenge of the twenty-first century is the same challenge we've always faced: to be better than we've been before.

Code Red

In 2017, University of Alberta virologist David Evans started buying DNA through the mail. Placing orders with commercial labs, he got genetic fragments—each about thirty thousand base pairs long—delivered to his doorstep. He used standard molecular techniques to stitch them back together. For less than $100,000, Evans and his team synthesized the horsepox virus, a close relative of smallpox, from mail order parts. His stated goal was vaccine research. Biosecurity experts saw a blueprint for terrorism. Evans's work proved just how easy it was to assemble a pandemic from scratch.

Smallpox was once humanity's most feared disease. With a 30 percent fatality rate, it killed three hundred million people in the twentieth century—more than both world wars combined. The campaign to eradicate it began in 1796, with Edward Jenner's first vaccine, and culminated in a global effort led by the WHO. On May 8, 1980, smallpox was declared eradicated—the first disease eliminated by human effort.

We may not be so lucky next time.

Since 2009, scientists have identified over nine hundred new viruses with pandemic potential. The Center for Global Development believes there's a 47–57 percent chance of another deadly pandemic before 2050.

And that's only natural outbreaks.

The tools to engineer life are cheap, fast, and, as David Evans discovered, readily available. DNA sequencing has dropped from $100 million per genome in 2001 to a few hundred dollars today. Gene-editing kits can be purchased online. CRISPR is in classrooms. Biology has been democratized. And, as Evans's work also highlights, the potential for engineered pathogens is a far greater concern. Rogue actors. Terrorist cells. The 5 percent of the population that psychologists identify as sociopathic. After five people died in the 2001 anthrax attacks, the US Postal Service spent over $100 million decontaminating their facilities. The CDC estimates that another anthrax attack could kill thirty-three thousand people for every one hundred thousand exposed and cost $26.2 billion.

Once again, accelerating technology is both the threat and the cure.

AI-driven biosensors have revolutionized point-of-care testing, offering real-time diagnostics from breath and blood. These same sensors could be deployed in airports, train stations, and high-traffic areas, detecting pathogens before they spread. Artificial intelligence and quantum computing are accelerating genetic analysis and shrinking vaccine timelines from years to weeks. One proposal links these developments into an early warning system that detects pathogens and designs defenses before the first symptoms appear.

Yet, technology isn't enough. We need global cooperation. The WHO's Global Health Emergency Corps is pushing for international collaboration on pandemic preparedness, initiatives like the Global Virome Project want to identify and catalog potential pandemics before they emerge, and SecureBio is taking an open source approach, creating global networks of scientists and citizens working together to monitor and prevent emerging diseases.

Is any of this happening fast enough? It took Evans and his team a few months to build their virus. That was almost a decade ago. Today's AI-powered tools can duplicate the feat in days.

So, no, none of this is happening fast enough. And the gap between what's needed and what's under way is widening by the day.

The Intelligence Explosion

Earlier, we met the "Godfather of AI," Geoffrey Hinton, and examined his pioneering work on neural networks.

In 2023, the Godfather quit.

Hinton left his job at Google and sounded the alarm. "It is hard to see how you can prevent the bad actors from using [AI] for bad things," he told *The New York Times*. A few weeks later, Hinton joined a who's who of AI pioneers, signing the Center for AI Safety's official statement about these dangers: "Mitigating the risk of extinction from AI should be a global priority alongside other societal-scale risks such as pandemics and nuclear war."

The proof is no longer in the pudding. Now it's in the doubling.

The raw computing power driving AI forward has been compounding every three months since 2022. That's Moore's law on steroids. In 2024, AI was a million times more powerful than it was in 2022. By 2030, the numbers experts toss about are nearly incomprehensible.

Maybe a thousand times more powerful. Maybe a million.

These same experts have been warning of the impact of this intelligence explosion. AI's an existential danger. A demon. A system that might replace 80 percent of the workforce. Yet at the front end of any major revolution, predictions tend to be binary—dystopia or utopia.

Typically, the real future unfolds along a middle path.

Consider BMW's experiment with robots and AI on their assembly lines. First, they replaced all their human workers with robots and AI. Productivity crashed. Then they reintroduced humans into the chain, using robots and AI to augment their skills. Productivity shot up 85 percent. This is the middle path that experts tend to ignore in their rush to capture headlines with the latest tech threat.

Or look at the legal industry, where AI-powered discovery tools haven't eliminated paralegals; they've created more demand for them. The pattern repeats across industries, and through time. In 1870, 70–80 percent of American workers were farmers. Today, it's 1.3 percent. Yet we didn't end up with 79 percent unemployment. Instead, we invented entirely new categories of work.

Or remember how the World Wide Web was going to gut the economy? For every job the internet eliminated, it created 2.6 new ones.

Yet job displacement and worker reskilling might be the front end of a much graver wave. We highlighted some of these concerns in the last section, but just to underscore: In 2023, a team of researchers created an AI system that could generate novel biological weapons, for under $2,000 and with publicly available tools. The same algorithms that may solve climate change can also be used to crash power grids. Rogue actors. Rogue AI. The news is not good.

The disinformation threat is equally real. In 2024, AI-generated robocalls mimicking President Biden's voice tried to suppress voter turnout in New Hampshire. Deepfake technology has experts struggling to distinguish the synthetic from the authentic. We're entering what Princeton professor Michael Graziano calls a "post-truth world," where seeing is no longer believing. Earlier, we talked about the critical importance of developing rigorous truth filters in an exponential world. Disinformation emphasizes that advice.

But the existential threat that is keeping us most awake has been keeping us awake for a very long time. In his 1872 novel *Erewhon*, British writer Samuel Butler imagined machines that could evolve, first toward consciousness, then toward intelligence exceeding our own, and finally toward rebellion. In the 1947 sci-fi novel *With Folded Hands*, author Jack Williamson gave us humanoid robots who do revolt, taking over the world under the guise of protecting humanity from itself. The notion of the AI apocalypse—where an AI actively destroys the world—is usually credited to Harlan Ellison's 1967 story "I Have No Mouth, and I Must Scream." And if you're looking for Terminator-style AI destruction, that would be Fred Saberhagen's *Berserker* series from 1967, which feature self-replicating AI war machines that wipe out all biological life.

So, nothing new under the sun, except progress.

Imagine an artificial intelligence a billion times smarter than Einstein. Not in a century. Not in fifty years. But by the 2030s. Earlier, when we said the numbers became unimaginable, this is what we were trying to imagine.

But unlike biological intelligence, artificial intelligence doesn't seem to plateau. Once AI reaches human-level intelligence, it can continue to improve its own code—and with no upper limit. Each improvement makes it better at making improvements. It's an intelligence explosion that explains why Stephen Hawking famously said,

"The rise of powerful AI will either be the best or the worst thing ever to happen to humanity. We do not yet know which."

This brings us back to Mo Gawdat and the "parenting problem." With AI, we are raising a new form of intelligence. Like children, AI systems learn from observing us and modeling our examples. Unlike children, AI learns at the speed of exponentials.

Task a superintelligent AI with solving global warming. Logical solution? Eliminate the species causing the problem. Ask it to end poverty? Maybe it decides to standardize living conditions at a subsistence level. These aren't just thought experiments; they're what AI safety researchers call *alignment problems*.

This is why companies like Anthropic are pioneering "constitutional AI" systems with built-in ethical constraints; DeepMind's "recursive reward modeling" is teaching AI systems to learn human values; and initiatives like the Future of Life Institute are pushing to pause development until better safety protocols are in place.

Neuroscience is even lending a hand. Researchers at the Allen Institute for Brain Science are mapping how human brains encode moral reasoning. The goal is to create AI systems that don't just mimic human intelligence but also our wisdom. Then again, it was human ingenuity that opened Pandora's box, so this, too, might be hubris.

And all of this is in the future. In the present, the biggest threat to our survival may be the energy required to keep the machines turned on. As AI grows more capable, it also grows more power hungry. Eric Schmidt recently warned that data centers could demand up to 96 gigawatts of new power by 2030—nearly the entire current US electrical capacity. "AI will drive an urgent need for *all* forms of energy," he said. And he's right. Each new breakthrough consumes more juice. We are creating minds that run on silicon but feed on terawatts. And unlike biology, these minds never sleep.

The response has been a global power scramble. Microsoft just signed a contract to restart the Three Mile Island Nuclear Generating Station by 2028. Google, Meta, Oracle, Amazon, and Nvidia are all inking long-term deals with start-ups in fusion power, SMRs (small nuclear reactors, a.k.a. "backyard nukes"), and generation IV nuclear reactors—some experimental, others already under regulatory review. Meanwhile, China added more solar in 2023 than the entire world did in 2022, and did it again in 2025. It's a planetary power race to keep intelligence accelerating. The big hope is that our AIs will help us redesign our planet's grid to feed our AIs—and sooner rather than later, because the meter is running.

As we close our examination of the downside of up and the very real dangers of acceleration, our big hope is that we've left you with hope. Yes, optimism can wobble in the face of this onslaught. Yes, challenge and overwhelm are the likely result. Many experts believe that our actions over the next decade will determine if we live to see the decade after that one. We don't want to undersell the nature of the threat, but we do want to remind you that this isn't the end of our story.

In part 3 of our survival guide, we'll lay out a blueprint for thriving in the age of abundance. We'll explore practical, tactical steps, using the knowledge of neurobiology and the tools of technology to upgrade our psychology and our biology, giving us a path toward prosperity that we can all travel together. And while change must start with each of us, it's our ability to cooperate at scale that ultimately determines our future. But one thing is certain, as Peter's son, Jet Diamandis, pointed out when he was thirteen years old: "Only when we manage the negative consequences of abundance and its related technologies can we truly use abundance to its fullest."

PART 3

THE AGE OF ABUNDANCE: A SURVIVAL GUIDE

Alone we can do little; together we can do so much.
—HELEN KELLER

CHAPTER SEVEN

Mind 2.0

Do you feel divine yet?

Or, after our whirlwind tour through the light and dark sides of abundance, do you feel overwhelmed by acceleration and adrift in a world you barely recognize?

If it's the latter, you're not alone. We feel it, too. Everyone we know feels it. And here's where all that disquiet becomes costly: The moment we start believing the future is happening *to* us rather than being created *by* us, we surrender agency. In formal language, we adopt an external locus of control. In less formal terms? A victim mindset.

An *internal* locus of control means we feel in charge of our lives and able to influence our destinies.

An *external* locus is the opposite. It's powerlessness. The sense that life is happening to us and there's nothing we can do to change our fate. This mindset creates a neurobiological problem.

The brain is an energy hog. It uses 25 percent of the body's energy at rest. Ask it to do something hard—like surf a tsunami of exponential change—and that number skyrockets. So, the brain's top priority is always to conserve energy.

When we adopt an external locus of control—when we decide AI is an unstoppable tidal wave crashing onto our heads, for example—we send the brain the signal, "You're not in control. Nothing you do will matter."

Thus, the brain does what it's designed to do. It powers down. No more creative solutions. Innovation ceases. Hibernation begins. The brain stops planning, stops preparing, stops reaching. Instead, it retreats into what psychologist Martin Seligman called "learned helplessness."

From a neurobiological perspective, this really is when all hope is lost.

Agency is more than a feeling. It's a foundation. The brain is a prediction engine, constantly scanning the world for familiar patterns, matching present inputs to past experiences to simulate likely outcomes. This is how it keeps us alive: by anticipating what comes next and budgeting energy accordingly.

But in a world moving this fast, the brain can't find familiar patterns. Our predictions keep failing. This triggers the salience network: stress, cortisol, norepinephrine. Eventually, the system overloads. Too much novelty, too much uncertainty, too little signal. Your brain throws up its hands and says: *No idea what's going on. Good luck out there.*

That's when helplessness is learned. It's the seed kernel of despair.

But it's not a new problem.

The myths tell us the old gods were a hot mess. Zeus threw tantrums disguised as thunderstorms. Hera turned jealousy into an Olympic sport. Dionysius made chaos his brand. If Spider-Man's maxim is "With great power comes great responsibility," the myths tell us that responsibility requires practice.

Even the gods had to learn foresight, restraint, and collaboration. Which brings us to a peculiar question: If humans now have godlike powers, how do we learn to wield them with godlike wisdom?

For all its arrogance, this question may end up the defining challenge of our era. Technological acceleration creates a neurobiological conundrum. Evolution optimized the human brain for a local and linear environment, not a global and exponential world. Linear progression we can handle. Exponential leaps? That's when the wheels come off. Abundance disrupts our mental models.

Recalibrating from linear to exponential thinking requires a mental shift as profound as the Enlightenment. Yet the Enlightenment was a cultural transformation that unfolded over centuries. Abundance is arriving in decades. This unprecedented pace demands unprecedented agility. This is the goal of part 3 of our survival guide.

In this final section, we arm you with science-based tools, techniques, and tactics to thrive in our amped-up environment. Our emphasis here is deliberate: Every intervention we explore is grounded in neurobiology. That's our lifeline. When the modern world moves faster than our prehistoric instincts evolved to handle, we can't keep pace by willpower alone. We need to work with the brain, not against it.

That work begins where perception begins: with frame, mindset, and bias—the brain's three filters for making sense of reality. Then we'll move into creativity, purpose, and flow—three of the most powerful levers we have for reclaiming agency and navigating the noise. And we know what you're thinking: *Fight back against an exponential explosion with mindset? Seriously?*

Yeah. Seriously.

For starters, there's no choice in the matter. Our psychology is our interface to our neurobiology. We have to upgrade our minds to keep pace with our world. It's the first step toward thriving.

Next, we have to use our retuned brains to learn to do something we've never really done before: cooperate at a global scale. This is the key point. AI is not happening *to* us. It's being created *by* us. This is true for every existential threat explored in this book. We're not rag dolls in raging winds of change. We are the raging winds of change. If we don't learn to collaborate at scale, we may not survive long enough to see what scale really looks like.

That's this chapter.

In the next chapter, neurotechnology enters our tale. We examine the exponential acceleration of brain-computer interfaces (BCIs), consciousness-raising technologies, and other upgrades designed to help human wetware keep pace with AI hardware, VR software, and reality, seriously augmented.

In chapter 9, we zoom out to confront a deeper question: Are we ready for a world of everything, everywhere, all the time? We'll examine the science of longevity and the strange, sweeping consequences of living a very long time in a post-scarcity world. Where does meaning come from? What happens to purpose? Once, these were philosophical questions. Today, they're survival skills.

But we're getting a little ahead of ourselves. Here, now, our education must start where all education starts—in the dark confines of the brain.

Cognitive Filters: Frame, Mindset, and Bias

Happiness, for all the hoopla, is an information-processing problem. The brain takes in eleven billion bits of information a second. How we filter and interpret this information determines the quality of our lives. So how does the brain sift and sort incoming data? That's the trio of mental skills known to psychologists as *frame*, *mindset*, and *bias*. These skills are a sieve for all incoming information, transforming chaos into coherence, and shaping our reality as a result. But to tune our brains for the reality of abundance, we need to recalibrate these filters for a world of overwhelming complexity.

In other words: Warp speed, Scotty.

Neuroscience provides a road map. Our prediction engine brain compares present scenarios to past events, operating like a biological Bayesian engine. When expectations fail to align with experiences, the brain generates a prediction error: a signal that we need to update our mental models. Here's where mindset, bias, and frame come into play. Mindset defines the structure of those mental models. Bias determines which details to emphasize and which to ignore. Frames provide context and convert perception into action. In an exponential world, we need to train all three of these filters to make a new kind of prediction.

Let's start with mindset.

A stable belief system that influences how we process experience, mindset is a top-down blueprint for the interpretation of information. This isn't positive thinking. This is a trainable tool that reduces threat perception and enhances creative problem-solving.

Details are useful.

The anterior cingulate cortex is the part of the brain that notices prediction errors. It uses this information to make decisions about how we make decisions. When there's uncertainty, the anterior cingulate cortex becomes logical and linear. It stops seeking innovative solutions. Under the threat of maybe, it demands the tried-and-true. But this strategy fails in the face of exponentials. A world of abundance requires an abundance of creativity. As Albert Einstein reminds us, "You cannot solve a problem with the same mind that created it."

Bias, by contrast, is a bottom-up process, and one of the brain's main methods for conserving energy. To deal with the data dump of daily life, the brain uses heuristics. These are the decision-making shortcuts—like common sense—that we discussed in chapter 1. Heuristics are designed to simplify problem-solving. But, as we also discussed, they don't always work as designed.

When heuristics fail, they become biases, reality-warping filters that prioritize speed over precision. Biases reduce uncertainty by streamlining decision-making, but this comes at the cost of accuracy.

Our negativity bias amplifies threats, a survival mechanism in days of yore. Today, it's why we think the apocalypse is nigh when our inbox is full. Confirmation bias entrenches existing beliefs. Status quo bias makes us resistant to change. Both conspire with loss aversion to favor the familiar over the novel. Then recency bias compounds these issues, causing us to overemphasize current events while underestimating long-term patterns.

Like mindset, biases are rooted in the drive to minimize uncertainty. They simplify the world, reducing cognitive load and increasing efficiency. But in the face of exponential change, those same shortcuts limit our adaptability, because ancient instincts aren't built for the speed of modern reality.

Frames, our last category, are the brain's lens for the real-time interpretation of events. More transient than mindsets or biases, and often imposed by external factors, such as language, culture, or context, frames change how we respond to a situation. Seeing a challenge as an opportunity alters the brain. Uncertainty diminishes, cognitive dissonance is reduced, and survival responses become creative and exploratory.

Is AI coming for your job?

Framing this as an existential threat triggers terror. Reframing it as an opportunity to reskill unlocks agency. Amazing are the powers of our minds.

When they work, mindset, bias, and frame help us navigate complexity with clarity. When they fail? The world becomes a hall of distorted mirrors. The good news is that these three filters are mutable. A host of interventions—reflection, mindfulness, flow, exposure to diverse perspectives—shift our cognitive shortcuts so our predilections match our predictions. We can tune these processes to adapt to a world that won't stop changing.

Or not.

You can keep relying on cognitive habits shaped for a different era, but they won't get you very far. That flash of confusion? That's your biology realizing its way out of its depth.

Where to Start

If you want to retune your brain's filters, there's a practical order to the process. Start with frame, move to mindset, then close with bias. These systems are interwoven, but their accessibility and their timing vary. Frames are conscious and immediate lenses we can shift in the moment. Mindsets run deeper. These semiconscious patterns influence how we interpret events over time. Biases operate the furthest below the surface. They are automatic shortcuts the brain uses to save energy. However, the interaction moves in both directions. A new frame can shape a mindset. A shifted mindset can loosen a bias.

Take *gratitude*.

When we feel grateful, we signal safety to the brain. This might not sound like much, but it undercuts the victim mindset, reduces background threat detection, and—over time—recalibrates our negativity bias. This reweighting matters.

The classic intervention takes five minutes. Write down three things for which you're grateful. Choose one and expand on it. Turn that into a paragraph. Feel the gratitude in your body. This somatic signal is the catalyst that shifts attention toward the positive, lowers stress, increases creativity, and boosts problem-solving. Once again, behold the powers of the brain.

And there are other useful tools for reframing—a whole suite of them.

One is *assumption questioning*. Our stress responses don't always begin in reality. We don't see what's happening; we see what we assume it means. Pausing to interrogate assumptions creates mental space for better choices. Ask: *What story am I telling myself right now? Is it true? What else might be true?*

Is that Godzilla? No. Just my parents calling to ask if I'm dating anyone seriously yet. Turns out not every spike in my threat response is an existential event.

Questions interrupt automatic interpretations. They open the door to more adaptive responses. Again, the reason is neurobiological.

In the brain, anxiety and excitement are triggered by the same chemical: norepinephrine. The difference between these emotions lies in interpretation. Reframing a challenge as a thrill and not a threat shifts how the brain allocates attention and energy. This is why a simple cognitive pivot—asking what else might be true—can turn fear into fuel.

Novelty exposure is another tool. Bias thrives on repetition. So, seek out new experiences, new perspectives, and new environments. This forces the brain to update old models, expanding cognitive flexibility and priming creativity. As with gratitude, the goal isn't to pretend the world is safe or that life is easy. The goal is to train your brain to respond with curiosity instead of panic.

This is the starting point. Reframing. Repatterning. Reclaiming the filters that shape your reality. If you want to thrive in a world that won't stop changing, you need a brain that can flex with dexterity.

The Creativity Compass

In the early 1980s, Pixar wasn't a film studio. It was a failing hardware company. Born inside Lucasfilm and later bought by Steve Jobs, the original Pixar team built the Pixar Image Computer—a sleek, powerful machine designed for medical imaging and scientific visualization. Nobody wanted it. Not hospitals. Not research labs. Not even George Lucas.

Pixar was bleeding cash. Their product wasn't selling. But they had an odd advantage: a small in-house team that made short, animated clips to demonstrate what their computer could do. While the tech flopped, the clips blew people's minds.

So Pixar pivoted. They gave up on hardware and bet everything on storytelling. What followed was a revolution. *Toy Story* became the first feature-length film made entirely with computers. It changed animation forever—and made Pixar one of the most successful studios in Hollywood history.

The Pixar pivot wasn't just a business move. It was a creative leap from silicon to storytelling. They went from optimizing a tool to reimagining its purpose. That's the power of creativity under pressure. It doesn't just solve problems. It removes blinders and rewires the possible.

Why does this matter here? Because frame, mindset, and bias filter the data we take in, but creativity is how we turn it into inspiration. Technically, creativity is the process of producing original ideas that are useful. Practically, it's a survival skill for accelerating environments that allows humans to adapt with dexterity, flexibility, and ingenuity, and all at once. Creativity helps us build new systems when the old ones collapse. It's how we navigate when the maps fail, or, like today, when we're completely off the map.

Creativity is different from everyday problem-solving. Neurologically, it's closer to a state of consciousness than a skill. In this creative state, three of the brain's major networks radically alter their behavior. Normally, the cognitive control network, which governs focus and filters out distraction, and the default mode network, the seat of mind-wandering and imagination, work in opposition. One goes up, the other goes down. This is what keeps us from daydreaming about our vacation in Greece while we're operating a jackhammer. In the brains of creatives, however, these networks coactivate and collaborate.

Additionally, the salience network—the toggle that switches attention between networks—becomes flexible and dynamic, directing traffic between those other two systems with increasing ease. Frontal lobe activity relaxes. Alpha and theta brain waves rise. Focus shifts, memory loosens, and ideas begin to recombine. It's the magic of network dynamics, a shift in neural configuration that unlocks an entirely different class of cognition.

And the effects are sweeping: better pattern recognition, faster learning, improved decision-making, greater psychological resilience, stronger immune function, and even increased longevity. But what matters most is that creativity underpins adaptive flexibility, which is the true core of resilience. Because it lets the brain shift strategies, perspectives, and behavior in real time, adaptive flexibility makes people harder to break. And when amplified at scale, it makes cultures more resilient. Companies that generate new solutions outperform those that cling to old systems. Nations that reward original thinking navigate cultural upheaval more deftly. Societies that invent new stories tend to survive the collapse of old myths.

Mihaly Csikszentmihalyi, the godfather of flow psychology, tied the cultivation of creativity to the success of civilization, writing in his book *Creativity*:

> It is possible that children who were more curious ran more risks. . . . But it is also probable that those human groups that learned to appreciate the curious children among them, and helped to protect

and reward them so that they could grow to maturity . . . were more successful than groups that ignored the potentially creative in their midst. If this is true, we are the descendants of ancestors who recognized the importance of novelty, protected those individuals who enjoyed being creative, and learned from them. Because they had among them individuals who enjoyed exploring and inventing, they were better prepared to face the unpredictable conditions that threatened their survival.

Creativity is the system that evolution designed for anti-fragility, and in our survival-of-the-speediest world, it's our compass. We built tools of mythic power: AI, robotics, synthetic biology, planetary-scale networks. Yet we're steering them with software tuned to life on the savanna. Logic helps, but lateral thinking, those strange intuitive leaps between distant ideas, is what lets us find the signal in the noise, replacing cognitive overload with clarity and choice.

In a world of artificial intelligence, lateral thinking is also our ultimate advantage. AI can remix ideas at speed, but it can't intuit what hasn't been imagined. It predicts, completes, and converges. It does not diverge unless it's hallucinating. Those wild leaps remain ours alone, and that's the frontier and the solution.

In education, rote memorization no longer drives success the way it once did, and creative problem-solving now tops every list of essential skills. In science, the real edge is no longer brute-force logic but the ability to bridge domains and ask better questions. In business, perpetual innovation has replaced strategic planning, forcing companies to build for change instead of protecting what already exists. And in life, the constant reinvention of self—call it the creative's dilemma—is no longer the exception. Gen Z'ers change jobs 134 percent faster than they did five years ago. The fallout shows up everywhere: in our careers and in our communities, where professional arcs have become scatterplots and identities must be retooled mid-flight. Stability is dead. Agility is king.

The good news is that creativity solves both sides of the challenge, giving us the cognitive flexibility we need to adapt and the motivation to stay in the game. Novel thinking lights up the brain's reward system, amplifying drive and boosting pattern recognition. That's when the shift happens. Now, creative ideas don't just appear—they cascade. One insight triggers another, perception widens, and ideas start to connect.

And creativity isn't fixed. It's trainable. Like a muscle, it strengthens under friction, through use, and in the face of challenge. Even fifteen minutes a day of writing, drawing, dancing, or designing starts to rewire the brain. Creativity microdosing, as this practice is known, widens perspective, sparks insight, and generalizes into other areas. It's cross-domain plasticity: faster learning, better insights, and the foundation of fruitful collaboration.

It is also essential. For survival.

When Arie de Geus, head of strategy for Royal Dutch Shell, studied the world's longest-living companies—some more than seven hundred years old—he found the ones that thrived through the chaos of the centuries had one trait in common: accelerated learning. The fastest learners survived, through wars, famines, plagues, even ice ages. De Geus called these organizations "living companies" because their longevity came from the biological capacity to regenerate in the face of change rather than defend the castle against all invaders.

Yet learning speed is driven by imaginative range. If you can't reframe or recombine ideas, you resort to instinct and habit. But if you can think differently on demand, that's how you bend without breaking. This is what resilience looks like at the cognitive level, and it's why creativity underpins every form of high performance, no matter the domain. Most crucially, in an era defined by disruption, creativity is how we steer at speed.

The Purpose of Purpose

In May of 1961, President John F. Kennedy made a promise so outrageous it bordered on fantasy: The United States would send someone to the moon before the end of the decade, and bring them back alive.

At the time, the United States had only put one astronaut into space. The flight lasted fifteen minutes. NASA didn't have a lunar module. Or a guidance computer. Not even the materials needed for a heat shield that could survive reentry.

Kennedy wasn't describing a plan. He was declaring a purpose.

This purpose became a gravitational force that pulled together physicists, engineers, chemists, computer scientists, and policymakers. It sparked the invention of the integrated circuit, the birth of satellite communications, and the creation of mission control. Over two hundred new technologies were built from scratch—not as the goal of the mission but as a byproduct of its purpose.

"We choose to go to the moon in this decade and do the other things," Kennedy said, "not because they are easy, but because they are hard." But that wasn't the full quote. He added: "This country was conquered by those who moved forward—and so will space."

That's the purpose of purpose. It reframes the impossible as inevitable by transforming constraint into creativity. This forces the brain to behave differently, becoming more focused, more robust, more dynamic. So why follow a section on creativity with one about purpose? If creativity is how we steer, purpose tells us where to go. It's the North Star of decision making.

Purpose changes the structure of how we think, which, at a neurobiological level, solves multiple problems at once. A clear mission recalibrates the filters that shape perception—mindset, frame, and bias—and it does all this at speed. Purpose also activates the brain's reward system, releasing dopamine, the neurochemical of drive. Push rock up hill. Repeat. It's dopamine that helps us play Sisyphus. When obstacles appear, purpose reframes them as opportunity. The brain stops asking, "Why bother?" and starts leaning toward "What's next?"

This shift is structural. Purpose moves us from anxiety to agency. It heightens activity in the prefrontal cortex, the seat of long-term planning and problem-solving, and suppresses activity in the amygdala, the brain's fear center. Threat detection becomes possibility generation, and opportunity follows.

Purpose transforms mindset as well, acting as a higher-order filter that shapes how we interpret the world. It tunes out noise and locks attention on target. This alignment boosts neuroplasticity. When we care about the outcome, the brain allocates more energy to learning and performance. Neural circuits reinforce themselves. Behavior becomes belief. That victim mindset gives way to adaptive agency.

Purpose also reshapes bias. Our cognitive shortcuts evolved to keep us alive in a world of scarcity; they were optimized for survival, not strategy. Purpose overrides outdated heuristics. Present bias gives way to long-term planning. Confirmation bias becomes curiosity. We stop scanning for threats and start seeking opportunity. This is why psychologists refer to purpose as *meta-adaptive*. It doesn't help us feel better. It helps us become better. People with purpose experience less stress, recover faster from setbacks, enjoy better physical health, and are more likely to access flow. In an exponential world, this makes purpose essential.

JFK's moonshot wasn't an isolated incident. It was a blueprint for collective ambition. A demonstration of how a bold intention, backed by commitment, constraint, and clarity, can mobilize creativity across domains. It's also a blueprint for solving the hard problems we now face, from climate change to rogue AI to the next pandemic. In a world moving this fast, purpose is the organizing principle that makes large-scale solutions possible.

Mindsets That Matter

Purpose gives us a reason to stay the course, aligning our goals, rewiring our perception, and activating the deeper circuitry of motivation. But in a volatile world, purpose needs reinforcement, which is where mindset reenters the story.

Mindsets are the operating systems that stabilize how we think and behave. They're not belief in a narrow sense; they're interpretive frameworks—durable enough to shape behavior across decades, yet flexible enough to update when facing accelerating change. When tuned properly, they stabilize frames, filter bias, optimize creativity, and reinforce purpose. Mindset mastery is the full tune-up. And in a world of information overload, it's also neural self-defense.

Here are the five mindsets that matter most.

The Curiosity Mindset

Curiosity is our foundational fuel. It's one of the brain's basic motivators, underpinned by dopamine and designed for discovery. That spark propels us to investigate, which drives insight and learning, forming the scaffold for passion and purpose. Curiosity also widens our search space, challenges assumptions, and keeps our neural networks primed for exploration.

To cultivate this mindset, Csikszentmihalyi suggests (again in his book *Creativity*), three simple steps:

1. Try to be surprised by something every day.
2. Try to surprise at least one person every day.
3. Write down each day what surprised you and how you surprised others.

This is all it takes. Once we remind the brain that it's been surprised, the pattern recognition system automatically starts hunting for the reason why. This fosters curiosity and drives creativity, and momentum builds from there.

The Abundance Mindset

Our brains evolved in a world of scarcity. They default to threat detection, loss aversion, and zero-sum thinking. But those instincts can misfire in a world of abundance.

An abundance mindset replaces fear with possibility. It stops us from trying to cut the same pie into thinner and thinner slices and teaches us how to get innovative and bake more pies. It's a mindset grounded in the idea that resources can multiply when the right technology is applied. This shift moves the brain from competing over dwindling reserves to collaborating to create new ones.

To shift from a scarcity to an abundance mindset, independent research from psychology, neuroscience, and behavioral economics converges on the same finding: reframing is the lever with the biggest lift. Whenever you start to feel resource strapped, whether you want more time, money, ideas, or opportunity, treat those needs as expandable through collaboration, learning, and iteration. They almost always are. This single move has the strongest impact. But to really lock in that impact, we have to rehearse it, reframing scarcity as abundance again and again, until the brain makes the adjustment automatically.

The Longevity Mindset

A longevity mindset starts with a simple fact: a host of converging technologies, from AI and CRISPR to gene therapies and cellular medicine, are about to add healthy decades to our lifespans. More critical is the follow-on realization that the most important action one can take to intercept this healthspan extension is to maintain peak health today. This, then, is the longevity mindset: the belief that lifespan is more malleable than we once imagined, and that the right technologies can meaningfully extend our healthy years. In this view, aging is an opportunity. What would you do with fifty more years? A hundred? A longevity mindset says these are questions that demand real answers.

Those answers have an impact. Stanford psychologist Alia Crum discovered that having a positive mindset toward aging—that is, believing your best days are ahead of you—translates into an extra seven and a half years of life. That's the same gain a heavy smoker gets by quitting. The data is clear: how we think about aging changes how we age.

To cultivate a longevity mindset, you need to think and act as if aging is a modifiable process and not an inevitable decline. But this demands more than optimism. It requires action: taking ownership of your health and staying abreast of emerging therapies. Just as the abundance mindset teaches us that technology can multiply scarce resources, the longevity mindset teaches us that time is no different—it's a resource being multiplied by technology. Act accordingly.

The Exponential Mindset

Our brains were built for a world of stasis. For most of human history, little changed between the generations. A great-great-grandfather lived roughly the same life as his great-great-granddaughter. Tools, customs, and roles were all passed down like genetic code. The future looked like the past, the past looked like the future, and that was the way things worked because that was the way things had always worked.

Those days are long gone. Forget about the differences between the generations. Today, the world can shift on a daily basis. Change compounds. A capability that seemed impossible on Monday is disrupting an industry by Friday, and the challenge is both speed and scale. We're wired to think the world is local and linear, but the reality is global and exponential. When our prediction engine brain can't match the map to the territory, overwhelm becomes our default.

An exponential mindset rewires these expectations. Since exponentials double on a regular basis, we need to reset our baseline. Expecting rapid jumps in progress that can be measured in orders of magnitude alters perception and trains the brain to anticipate nonlinear leaps rather than rely on incremental logic. This recalibration is the foundation of an exponential mindset.

Another way to solidify this foundation is to forecast the unexpected. In an exponential world, the only constant isn't just change; it's acceleration. When trying to track a technology into the future, always start by asking how fast it doubles in power. Is it twelve months? Eighteen? Then project forward a few doublings, anticipating that the cycle shortens over time. This trains the brain to stay ahead of the curve and act accordingly.

The Moonshot Mindset

The moonshot mindset is a way of thinking that treats the impossible as achievable. By reframing the ridiculously difficult as the potentially doable, we open ourselves to new possibilities. This uses curiosity to prime the brain for exploration, amplifying creativity and unlocking the extraordinary.

The popular term for this process is the Bannister effect, named after Roger Bannister, who was the first person to run a sub–four-minute mile. It refers to the fact that after Bannister broke that record—long considered impossible—other runners followed in rapid succession. Yet the actual feat did not change. Running a sub–four-minute mile still required running a sub–four-minute mile. All that shifted was the mental frame that people built around the goal. This shift is the key to success. Once we see the impossible as possible, the brain starts hunting for the mechanics. What would make this feat doable? How could you chunk the problem into small steps? Where to start? What comes next? Pretty soon, what felt like a wall becomes a navigable path into previously uncharted territory.

We first introduced the term "moonshots" in our 2015 book *BOLD*, inspired by Astro Teller, whose title at Google X is "Captain of Moonshots." Teller's insight was that it's often easier to achieve a 10x improvement than a 10 percent gain. Why? Because incremental thinking keeps you trapped in the assumptions of the past. Moonshots force you to throw out old ideas and rethink the problem from first principles. They arise at the intersection of huge problems, exponential technologies, and the solutions these technologies provide.

A 10x goal breaks the frame. It demands you abandon conventional wisdom and rev up your creative potential. The ambition of the goal creates urgency, amplifies persistence, and inspires collaboration. Thus, when confronting a problem, think 10x, not 10 percent. Ask Teller's question: What would an improvement of this magnitude require?

Moonshot thinking isn't reckless optimism. Teller calls it "enthusiastic skepticism." This mindset encourages you to hunt for both the limits in your thinking and the flaws in your ideas. It rewires ambition by stretching belief. Once the brain treats limits as temporary conditions rather than fixed boundaries, entirely new lines of action come into view.

Mindset Mastery

We've been working our way through a series of mindset shifts, but the reality is that all these changes interlock. Curiosity fuels exploration. Abundance reframes scarcity. An exponential mindset aligns us with the true pace of progress. A longevity mindset expands our limits and our ambition. And a moonshot mindset pushes us to aim higher. Together, they form the necessary neural tool kit for thriving in an age of acceleration, giving us the cognitive stability to hold steady on our purpose, the flexibility to adapt to challenges, and the psychological tools required to chase bold goals.

But mindsets are internal. Thriving is external. It requires action. And the most reliable way to transform mindsets into action is flow: a peak state of total absorption where we perform at our best and feel our best. In flow, the brain shifts from analysis—where mindset, bias, and frame dominate—into action. Intangible ideas gain physical momentum. Purpose becomes execution. Captain, we are cleared for takeoff.

208 Seconds Before Impact

On January 15, 2009, US Airways Flight 1549 departed New York's LaGuardia Airport. Bound for Charlotte, North Carolina. Sunny skies and a light breeze. Two hours of flight time and they were home.

They were not going home.

Ninety seconds later, the Airbus flew into a flock of geese. The birds got sucked into the engines. Both failed at once. Later, words like *catastrophic* and *unprecedented* were thrown around. The plane was three thousand feet above one of the most densely populated cities on Earth. With no power, no altitude to spare, and 155 souls on board, Captain Chesley "Sully" Sullenberger had 208 seconds before impact.

Sully didn't panic. He didn't freeze. He didn't even raise his voice. He considered returning to LaGuardia. He considered dashing to nearby Teterboro Airport in New Jersey. But the math didn't add up. He couldn't make either destination. Instead, Sully chose to do the impossible: execute a water landing in the middle of the Hudson River.

Later, when asked how he made that decision, Sully said, "It was as if time slowed down. I was able to focus clearly and block everything else out. My training took over. It was automatic, but deliberate."

That's flow—an optimal state of consciousness where action and awareness merge, time distorts, and decision-making becomes seamless. In Sully's case, it turned an aviation disaster into the "Miracle on the Hudson."

While Sully's miracle may be exceptional, the state of consciousness that enabled it is not. Flow is the tool evolution devised to help the brain perform at its best, especially in crisis situations. Yet the state is available to all of us, nearly all the time. Flow is hardwired, a built-in feature of being human, and, even better, it's a trainable skill. And in a world where the crisis of high-speed, high-stakes decision-making is every day and all the time, training this skill may be the only way to keep the plane in the air.

In flow, the brain processes information more quickly, more completely, with far greater accuracy, and with far less friction. Patterns that normally sit outside of conscious awareness come into view, so problem-solving occurs without hesitation and with fewer errors. The feeling is one of "effortless effort," every decision, every action, *flows* seamlessly, perfectly, to the next. You know what to do and you do it, no questions, no hesitation. That's the origin story for the state's name. Flow feels flowy. It's a phenomenological description of the experience itself.

Flow changes how the brain processes complexity, which is a polite word for the exponential blitzkrieg known as Tuesday. In flow, the prefrontal cortex, the brain's hub for self-monitoring, quiets down, and the drag of over-analysis falls away. We make quicker, more intuitive, and more inspired decisions. And then we act on them. Thought and action collapse into a single motion, precise yet, should conditions change, infinitely flexible.

This is why flow is a critical tool for problem-solving in today's world. In normal states of consciousness, the brain processes information habitually, and sequentially, using old ideas to solve new problems. This strategy works fine in the face of logical and linear challenges, but in a global and exponential world, the result is a cognitive bottleneck, which is a technical way of saying, "Big iceberg, dead ahead."

Flow bypasses this constraint. In the state, we parallel process information across multiple brain networks, so rapid insights, later connections, and intuitive breakthroughs arise with surprising regularity. This is why elite performers in every field—from Silicon Valley founders to Special Forces operators—spend their lives training for flow. It's the biology that allows us to thrive amid chaos.

Chaos means uncertainty in us and volatility in the world around us. Together, they form a potent stress load. Chronic exposure produces hypervigilance, erodes neural plasticity, and weakens executive function. In the worst cases, it leads to depression, burnout, and PTSD.

And none of this sounds like thriving.

Flow counteracts the damage. The state stabilizes mood, restores cognitive flexibility, and supports long-term well-being. It's also the point where every element in this chapter comes together. Purpose sets direction. Mindset determines how we interpret what we see. Bias shapes what enters awareness. Creativity gives us the ability to adapt. And flow is the mechanism that fuses these capacities into a single, coherent mode of action.

The Flow Consensus

In researching this book, we spent a lot of time talking to experts across a dozen fields: AI, biotech, genetics, robotics, neuroscience, psychology, and more. While their expertise spanned wildly different domains, a rare consensus emerged. And if you've ever spent time around experts, you know—a consensus sighting is rarer than Bigfoot.

The opinion these people shared was about who would thrive in the twenty-first century. Across the boards, everybody agreed: Creative leaders who know how to drop into flow and collaborate, with each other—and with AI—will own the future. Full stop.

Why? Because flow amplifies lateral thinking, or the ability to make unexpected connections between unrelated ideas. This is a skill that AI lacks. The nine-dot problem is the classic example. Imagine nine dots in a three-by-three box. The challenge is to connect all nine dots with four straight lines, without lifting your pencil from the paper, in ten minutes or less.

Assumptions take most people down.

People fail to solve the nine-dot problem because they assume the lines must stay inside the box. But to solve this puzzle, you must think outside the box. Literally. Connecting those dots requires extending your lines outside the square they form. This leap beyond logic is lateral thinking.

Humans are gifted lateral thinkers. And humans in flow? Extraordinarily gifted—this is the more important point.

Let's go back to the nine-dot problem. Under normal conditions, this problem stumps most people. Fewer than 5 percent of the population gets it right. But in a University of Sydney study, when researchers used transcranial magnetic stimulation to push people into flow, they were 58 percent more likely to connect those dots. That's a staggering increase in lateral thinking.

And here's the point: Lateral thinking is a skill that AI lacks. Large language models are convergent problem-solvers that utilize deductive reasoning to match like with like. This lets AI excel at efficiency, speed, and raw computation, but wild leaps between weird ideas is where the technology hits a wall.

When humans in flow collaborate with one another and with AI, something new emerges. Precision meets imagination. Logic meets lateral insight. This is the real story of the future of work. It's not human versus machine. It's humans in flow collaborating with machines.

Perhaps we've been telling the tale sideways. In the near future, will AI steal your job? Maybe. But a creative leader who knows how to drop into flow, amplify lateral thinking, and collaborate with AI? Yes, *they* will definitely steal your job.

So, final question: How do we train for flow?

Ticket to Flow

In 1890, William James published *The Principles of Psychology*, a two-volume, thousand-page masterwork that founded modern psychology. Dense, rigorous, written for researchers, it introduced a new way of thinking about the mind and brain.

In 1899, James returned to the subject with *Talks to Teachers on Psychology: And to Students on Some of Life's Ideals*. This time his audience wasn't scientists; it was students, teachers, and anyone struggling to stay sane amid the whirlwind of the industrial revolution. To help people deal with the upheaval, he condensed the wisdom of his thousand-page textbook into a simple directive: "The great thing then, in all education, is to make our nervous system our ally and not our 'enemy.'"

James meant it literally. Habits, attention, emotions were survival tools in a century speeding beyond comprehension. Master your nervous system, he argued, and you master the chaos.

In this chapter, we've been following James's advice. It's been a blueprint for rewiring the brain in an age of exponentials. Yet our protocols face the same issue as James's advice: They feel profoundly underwhelming in the face of the overwhelmingly real. The labor market is about to be disrupted by millions of autonomous robots, and the best way to meet that future is with mindsets, frame, and flow?

There's no other option. If we want to surf the tsunami of today, the brain is the only interface available. And the brain in flow is our best chance at a high-performing interface. A century after James urged us to make our nervous system our ally, we finally know how. Flow is that alliance in action.

To train flow, the most important detail is the most straightforward: flow follows focus. The state can only arise when all our attention is locked on the task at hand and in the present moment.

There are tools to help with this job. Flow states have triggers. These are preconditions that lead to more flow. There are twenty-eight in total. Twelve are triggers for individual flow. Sixteen are for group flow, the shared, collective version of the state. But what do they all have in common? They all amplify attention and drive focus into the now.

The place to start with is *complete concentration*, arguably the state's most important trigger. Distraction is the enemy of focus, so when you want to drop into flow, practice distraction management. Power down your phone. No email. No notifications. No messages. No social media alerts. Aim for 90-minute deep-work blocks. The brain has a 90–110 minute awake-and-alert cycle that matches our 90–110 minute sleep cycle. Giving the brain this much time to stay on task allows biology to work for us rather than against us.

Remember, most flow-shattering distractions come from friends, family, coworkers, and bosses—so have your conversations in advance. All these people want more of your time. Flow amplifies productivity (a 500-percent increase in a McKinsey study), which means you get time back by first making time for flow. This means the people who want more of your attention can have more of it, but only if they leave you in peace for a while.

The next trigger to layer in is clear goals. The brain is a goal-directed system; give it a specific mission and it's built to stay on task. Clear goals are nothing more than a well-crafted to-do list that tells the brain where to aim its attention. When we don't know what we're trying to accomplish, we burn time and energy trying to figure it out. Clear goals eliminate uncertainty. If you're an author, don't set out to write five hundred words in your new book today. Set out to write five hundred words that tell the story of your character arriving at the airport, moving through security, and boarding a flight—that's the level of precision we're looking for.

If clear goals define where we go, the challenge-skills balance is the trigger that describes what we do once we get there. Often called *the golden rule of flow*, the challenge-skills balance is another straightforward idea. Flow follows focus. We pay maximum attention to the task at hand when the challenge of that task slightly exceeds our skill set. You want to stretch but not snap. If the challenge is too easy, attention flags, and boredom takes over. If it's too hard, anxiety pushes us toward overwhelm. Inside the challenge-skills sweet spot, attention stays locked on target, and flow becomes the direct result.

By combining flow with the other tools in this chapter—layer in mindset, reframe for growth, obliterate limiting biases, anchor to purpose, and amplify with creativity, we stack the entire cognitive deck in our favor. Attention, emotion, and motivation all fall into alignment. Our nervous system stops behaving like our adversary. What we gain is adaptive flexibility and expanded cognition—the neurobiological footing needed to navigate an abundant world. Jump to light speed? No problem.

CHAPTER EIGHT

The Androids Are Us

The previous chapter examined ways to train the brain to thrive in a world of abundance. This chapter sticks with that theme, but switches focus to the other side of the equation: from psychological interventions that help us keep pace with exponentials to technological enhancements that do the same.

Surfing waves of accelerating change and not drowning along the way requires a new mental architecture. Since evolution moves at the pace of generation, and exponentials double in months, our biology can't keep up. We need technology to absorb the responsibility.

First, a little history is helpful.

The cognitive skills encountered in the previous chapter are the result of a two-decade revolution in neuroscience that began when President George H. W. Bush declared the 1990s "the Decade of the Brain." During this period, research dollars flooded into the field. The tools for peering inside the brain rode the same exponential curves powering the rest of this book. Room-size imaging machines shrunk to pocket-size wonders, while the computational power needed to analyze the data collected by these machines rode Moore's law right into the App Store. This convergence birthed a new generation of neurotechnology—or what could be called *brain-enhancing, consciousness-raising technology.*

Brain enhancement is for speed. Consciousness-raising is for scale. And the combination forms the foundation of our new mental architecture. Speed provides the faster data processing needed to outpace information overload and thrive amid uncertainty. It also allows us to sync up with AI, coupling human creativity with machine intelligence for mutual benefit.

Scale is both internal and external. Internally, the conscious mind has limited RAM. Working memory taps out after being loaded with three or four ideas. Since AI doesn't suffer these limits, keeping pace requires new ways to expand our limited storage capacity and enhance our recall. Externally, solving the existential threats examined in chapter 6 demands global cooperation, which is something of a tall order. Consciousness-raising means using neuroscience to unlock expansive altered states once reserved for mystics: deep flow, collective ecstasy, even the oneness with everything that comes from enlightenment. These states are the very technologies that the brain uses to foster and enhance cooperation. They spark compassion, expanding empathy and dissolving the boundaries between us. When technologically enabled—for greater precision and control—they form the cognitive infrastructure for large-scale collaboration, making them our best shot at neutralizing existential risk.

This chapter explores the evolution of the neurotech revolution, from early meditation apps and neurofeedback wearables to brain-computer interfaces, customizable psychedelics, and enlightenment gamification. Our goal remains the same: to tune the brain to keep pace with an accelerating world without losing our minds in the process.

The Birth of Consciousness Hacking

In the early 2000s, before consumer neurotech was even a category, Canadian artist and fashion designer Ariel Garten found herself drawn to the inner workings of the mind. She had studied both biology and psychology (and took neuroscience classes) at the University of Toronto, and soon began making art out of the brain's rhythms, trying to give an external form to our internal patterns. The result only ignited her curiosity more, sparking a collaboration with wearable-computing pioneer Steve Mann. The duo extended Garten's earlier work, and began experimenting with neuro-art, ways to turn brain waves into music and visuals. Her goal was to give people direct, everyday access to their own mental states.

Garten's timing was fortuitous. Just as she began making brain art, meditation was bounding out of monasteries and into the mainstream, riding that ultimate bucking bronco: the smartphone. Once these esoteric practices were rendered as code, they did what everything translated into ones and zeroes does: They hopped on the back of Moore's Law and began accelerating exponentially.

The first wave of brain-enhancing, consciousness-raising consumer neurotech was the result. In 2012, the pair of former Buddhist monks who founded Headspace (in 2010) launched their first meditation app. Their success pulled other companies into the fray. Next came Calm, then Insight Timer, then the floodgates opened. All these apps wrapped traditional practices in sleek, digital interfaces, transforming once secret techniques for spiritual advancement into widely available tools for performance optimization.

The technologies kept coming because the research on meditation kept piling up. At the same time these apps began hitting the mainstream, scientific studies about the benefits of mindfulness were flooding into research journals. Focused meditation decreases stress and improves attention. Loving-kindness practices enhance well-being and extend longevity. Vipassana techniques increase creativity and stabilize emotions. Health benefits, mood benefits, mind benefits. When it comes to optimizing the brain, it turns out that ancient wisdom had plenty to offer.

Scientifically, 2005 was the inflection point. That was when a team led by Harvard's Sara Lazar published a study in *NeuroReport* showing that mindfulness increased the thickness of the prefrontal cortex and hippocampus, cortical areas associated with attention, memory, and emotional regulation. It was the first proof that meditation physically changed the brain, and a finding transformed mindfulness from a spiritual discipline into a trainable skill, setting the stage for everything that followed.

The first wave of mindfulness technology built on this research, but it didn't build far. Those early meditation apps could guide people through exotic practices, yet subjective experience remained their only measure of progress. Users had to trust these practices worked their magic without any real-world confirmation. Is my focus sharpening? Am I really meditating? Or am I just sitting here with my eyes closed? Because it kind of feels like I'm just sitting here with my eyes closed.

The next step was quantification.

Self-tracking devices brought metrics to mental training. Fitbit, the Oura ring, and the WHOOP band turned sleep, heart rate variability, and stress recovery into numbers that could be optimized. At first, these wearables were for physical fitness. As data rolled in, it became clear that mental performance was equally measurable. Recovery scores correlate with productivity, deep sleep with memory retention, HRV with stress resilience. These devices became a gateway to a deeper understanding of cognitive function—but again, not that deep.

Wearables tracked the body. They measured the physiological aftershocks of meditation: lower resting heart rate, improved sleep. Not the mind itself.

Ariel Garten bridged that gap.

In 2009, she co-founded InteraXon with the mission to bring EEG out of the lab and into everyday life. The Muse headband was born out of the desire to do what the first wave of meditation apps could not: provide real-time neurofeedback on mental progress. Muse blended clinical-grade EEG with a futuristic headband straight out of *Tron*. It kicked the neurotech uprising into a higher gear both because it worked well and because it looked the part: one of the first devices of the twenty-first century that actually felt like it was from the twenty-first century.

Muse detects shifts between mental states by measuring changes in electrical activity in the brain. The experience is interactive. When a user meditates, the headband translates brain waves into auditory cues. A calm mind sounds like gentle waves. A distracted mind produces stormy weather. Instead of guessing whether you're actually meditating, you now have a direct window into neural experience. It's mindfulness with a feedback loop.

That feedback loop is critical for twenty-first century thriving. Mindfulness is a neurological workout that strengthens attention, quiets the amygdala, and steadies emotion in ways that sharpen many of the skills described in the previous chapter. As these circuits adapt, curiosity expands, creativity rises, and the fear-based shortcuts the brain likes to rely on begin to recede, making more room for reframing and choice. And because meditation trains the systems responsbile for sustained, deliberate focus, it reinforces the same underlying neural pathways that support flow.

Muse was the right tech at the right time. The stress of modernity primed the market for meditation, the meditation boom expanded the market for neurotech, and the wearables boom added in a hunger for qualification. Then Muse brought data to that party. Within a few years, the company had more than half a million users and one of the largest databases of meditation-related brain activity ever recorded. "Ten years ago," Garten explained in an interview with the authors, "the idea that you could steer consciousness reliably via a wearable technology was still science futurism. Now, we've not only got those wearables, we have the data to prove their effectiveness."

Muse's impact catalyzed a second wave of neurotech. Other EEG headsets entered the market, targeting a wider array of cognitive skills. NeoRhythm focused on brain wave entrainment. Emotiv developed tools for emotional regulation. Vital Neuro wove neurofeedback into the music so listening itself guides the brain into a more focused and relaxed state. Then OpenBCI gave us the hacker-friendly version of Muse built for researchers, while Halo Neuroscience expanded beyond EEG by using transcranial stimulation to enhance motor learning.

Finally, as neurotech's frontier advanced, more radical players entered the scene: Neurable, NextMind—it was quite a party.

Sure, there were issues. The signal-to-noise ratio of early EEG headsets was less than ideal. Devices struggled with interference from hair, muscle movement, and external electronic signals. Even so, the consciousness-raising trajectory was clear. The first wave was awareness: Mindfulness apps and fitness wearables made us aware of our cognitive states. The second wave was steering, which arrived when EEG headsets gave users interactive control over their minds. The third wave—the likes of Vital Neuro and Muse 2.0—solved those earlier problems and expanded from meditation into sleep, focus, and the full suite of cognitive enhancement.

Now, the field is entering its fourth wave: AI or, more specifically, the fusion of AI and neurotechnology. Instead of tracking and tuning cognitive performance, next-generation tools enhance it in real time. Human-AI collaboration is the new frontier. We're merging mind and machine.

Enter the centaur.

The Rise of the Machines

The early waves of neurotech were basic training, our first attempt to use brain-based tools to steady ourselves in the accelerating noise of modern life. Now AI is in the picture. It's amplifying the noise for certain, but it also boosts the neurotech we've developed to handle that pressure. It's a feedback loop between mind and machine that is starting to enable a new level of cognition. This is far beyond better sleep scores and smarter wearables. This next evolution is the full integration. Machines that understand minds. Minds enhanced by machines.

Call it the law of unintended consequences, hard at work.

Humans invent a new technology to solve an old problem, but those inventions create new problems. The internet solved the issue of communication at scale. It also gave rise to fractured attention spans and a global mental health crisis that, in turn, became rocket fuel for the accelerating field of neurotechnology.

First, we needed meditation apps to help undo the damage done by the modern world. Then came biohacking tools to improve our overall performance. The combination became a new way to navigate the growing complexity of our daily lives. Now, with AI, complexity is growing more complex, acceleration is accelerating, and we're struggling to stay ahead of the very curves we're creating. Yet this isn't a death match. It's not humans versus AI, or at least, not if we want to win. The solution lies in collaboration: humans plus AI—a hybrid-model known as the *centaur*.

The term comes from chess. In 1997, IBM's Deep Blue supercomputer defeated grandmaster Garry Kasparov in a six-game match. Afterward, Kasparov flipped the script—if you can't beat them, join them—and introduced the sport of "advanced chess," where humans partner with AI in a new form of cross-species collaboration. These human-AI teams were called *centaurs*, a nod to the half-human, half-horse creatures of mythology. In modern terms, centaurs combine human creative intuition and expertise with AI's raw computational power. And with stunning results.

In chess, for nearly a decade—before large language models (LLMs) rewrote the rules—centaurs outperformed both human grandmasters and solo machines. Kasparov's experiment proved that, when harnessed correctly, AI can amp up human intelligence to unprecedented levels, at least in tightly structured arenas like chess.

A decade later, that amplification left the chessboard and moved into just about every domain imaginable. A 2024 McKinsey study found that 72 percent of global organizations use AI to perform at least one core business function. That's nearly three-quarters of the planet. Yet while artificial intelligence races ahead, our ability to keep pace is lagging behind.

If the goal is better human-AI collaboration, we need to tune up our half of the equation, which brings us back to flow. We already saw how flow can amplify lateral thinking to maximize the benefits of human-AI collaboration. But the advantages go farther than creativity. By increasing the brain's information processing capacity, flow can help us keep pace with the velocity of our sped-up world.

To capitalize on this advantage, we need to design AI systems that produce and sustain optimal performance. Four key principles matter most.

Seamless interface design is the first. Flow arises when concentration is high and cognitive load is low. Intuitive, distraction-free design blocks interruptions and allows users to zero in on the task at hand. Google's minimalist search bar crushed AOL's clunky interface because it enabled deep work. The same principle applies to AI interfaces. Less is more.

Immediate feedback comes next. Real-time insights sharpen focus, amplify learning, and keep attention locked so the brain doesn't have to wonder about how to improve performance. It already knows. And that certainly tightens focus and drives flow. Poorly designed AI systems do the opposite, overwhelming users with information and distraction and pulling them out of the zone.

The Strava app is an example. Early fitness trackers provided post-performance metrics. You could see how you did only after the run or ride was over. This wasn't bad for tracking progress, but it didn't improve performance. Enter Strava. Now athletes had live metrics about pace, elevation, and race rank. This instant feedback amped up competition, which amped up motivation, which amped up focus, which amped up effort, and the whole stack pushed runners into flow. By gamifying feedback, Strava enhanced performance. AI interfaces need to follow similar principles.

Personalization and adaptability follow from here. One-size-fits-all AI won't work for flow. Twenty-first century centaur systems must adjust to a user's skill level. The goal should be to keep the user in the high-flow range, with challenges neither too easy nor too hard. Great video games do this automatically. Every AI system should follow suit.

Autonomy and mastery are our final considerations. Both are focus-tightening flow triggers. Autonomy and attention are coupled systems. When we're driving the bus, we pay way more attention to where the bus is going. We also like getting better at the skills it takes to drive that bus (i.e., mastery) for the simple reason that this improves our chances of surviving the trip. For these reasons, autonomy and mastery tighten focus and drive the brain toward flow. Centaur systems built on this model will naturally encourage exploration and experimentation (autonomy) as well as the ongoing skills development that fuels mastery.

These four principles give us a road map for designing the centaurs of tomorrow, but they're only half the picture. So far, all our attention has been on individual performance. Yet AI is changing more than how individuals work and flow; it's reshaping how we work and flow together. We don't live our lives in isolation, which is why the next AI-enhanced neurotech revolution is focused on team performance.

Syneurgy

Group flow is the shared, collective version of a flow state, defined as a team performing at their best. It shows up in a fourth-quarter comeback, when everybody is making that extra pass and nobody can miss; in the hive-mind energy of a great business meeting, where ideas are flying around the room; and in the effortless coordination of a surgical team, when the operation proceeds like a ballet.

It was psychologist Keith Sawyer who put group flow on the performance map. He spent over a decade studying the improv comedy troupe Second City Television, using audience laughter as a measure of group flow, in order to decode the phenomenon and identify its triggers (explored in his book *Group Genius*). He found ten triggers for group flow, a list that has since been expanded to twelve, but what they have in common is their ability to drive attention into the now. Unlike individual triggers, which marshal individual attention, group flow's triggers get everybody on the team focused on the exact same details.

Over the past decade, neuroscientists built on this foundation, identifying the neural mechanisms behind the skills that make group flow possible—collaboration, cooperation, shared attention, psychological safety, and more. This work began in the field of coordination dynamics, which explores how brains align in time, and continued as neurobiologists deciphered the chemistry of social bonding, from oxytocin's role in trust to dopamine's role in love. Social neuroscientists then learned to map and measure synchrony, entrainment, and inter-brain coherence. All of this culminated in a landmark 2021 study in which Caltech neuroscientist Mohammad Shehata used hyperscanning EEG to show that team flow has a unique brain signature: increased beta-gamma activity in the left middle temporal cortex that directly correlates to a significant increase in inter-brain synchrony, heightened information processing, and improved team performance.

As a result of all this work, AI tools are now being designed to target group flow's triggers, and GitHub's Copilot is an early example. It reduces cognitive load and heightens concentration with real-time code suggestions, and it tunes the challenge-skills balance by calibrating output to each coder's skill level. The platform isn't fully optimized for group flow, but it's a hint of what might be possible.

The next step arrived in 2021, when Dr. Michael Mannino, chief science officer for the Flow Research Collective (Steven's organization), teamed up with Erwin Valencia, an expert in leadership and team development, to co-found Syneurgy (that's *synergy* and *neuron* combined) to build the first AI-powered system to engineer group flow at scale. The concept is simple: track brain synchrony, feed it back into the system, and let the AI tune itself and the group. By monitoring everything from micro-expressions and speech patterns to heart rate and blood flow, the platform measures markers for shared attention and maps trust and psychological safety. Then it nudges, gently, with subtle cues, to help teams course-correct in real time. Syneurgy marks the first wave of AI-powered social neurotech designed to maximize collective attention. It aligns brains, strengthens synchrony, and drives group flow.

The system functions as both a tuning fork and a mirror held up to the room. Teams can see how aligned or misaligned they are in the moment and make changes on the fly. The outcome: cooperation flourishes. In a pilot study, Syneurgy found that synchronized teams can be up to three times more productive, see a 19 percent boost in overall performance, and have a 15 percent increase in group flow.

Syneurgy represents a much-needed development in our attempt to mitigate the existential threats created by exponential technology, all of which demand levels of cooperation and collective action that have so far eluded us. Consider *communitas*, the technical term for group flow at scale. When you're at a rock concert, and the entire crowd is clapping in synch, and everyone has merged with the music—that's communitas. It's thousands of entrained minds acting as one and the neural foundation of large-scale cooperation. What happens when the next wave of neurotechnology enables thousands of minds to align around a common goal? When synchrony can spread across movements, networks, and nations? Communitas suggests a new kind of cognition: distributed, dynamic, and deeply suited to solving the problems of the modern world.

The Dawn of Brain Augmentation

In 2005, Ray Kurzweil made one of the boldest predictions in tech history: By 2035, we would jack our brains directly into the internet. In his book *The Singularity Is Near*, Kurzweil claimed that exponential advances in AI, nanotech, and neuroscience would soon converge, unlocking brain-computer interfaces (BCIs) that would give us seamless access to the cloud. Bidirectional. High bandwidth. Tune in, turn on, surf the Matrix.

BCIs, in Kurzweil's opinion, gave us our best chance for keeping pace with AI, an extension of Kasparov's *if you can't beat the machines, merge with the machines* philosophy. If social neurotech was the first wave of AI helping humans cooperate—with each other and with machines—then BCIs are the brave new frontier, one with a very long history.

In the 1940s and '50s, studies on brain waves hinted at the possibility that these electrical patterns could be read, interpreted, and perhaps understood. By the early 1970s, researchers were capturing neural signals via EEG and translating them into computer commands. Twenty years later, quadriplegics began to be able to control computer cursors with their thoughts, and amputees started to gain control of their robotic limbs.

All this groundwork led to a turning point in 2000, when BCIs became headline news. Steven was in the room when Dr. William Dobelle turned on the world's first artificial vision implant. Two days later, the patient, who had been blind in both eyes for over twenty years, was driving a car around a parking lot. The story made the cover of *Wired* magazine, and for good reason. It was a miracle of biblical proportions: The blind could see.

Unfortunately, not for long.

The implant worked for a while, but it was a one-off experiment and equipment beta-test, and within a few months the technology faltered. Before it could be upgraded, the inventor died, the secrets of the implant went with him, and the patient was left with a useless hunk of metal inside their skull.

All of this is to say, in 2005, when Kurzweil made his bold prediction, BCIs were just that: a bold prediction. Within a decade, exponential technology turned prophecy into progress. The most famous example was Elon Musk's 2016 founding of Neuralink to develop high-bandwidth BCIs so humans could keep pace with AI. Neuralink's near-term goal is health focused: restoring sight, treating paralysis, and repairing spinal cord injuries. Their long game is Kurzweil's prophecy: merge mind with machine. Or, as Musk said in a recent interview: "Let's give people superpowers."

To make good on these promises, Neuralink started by developing flexible electrodes that do less damage than traditional stiff metal arrays. The electrodes integrate directly with brain tissue—1,024 channels packed into a device the size of a quarter. They also created a robotic neurosurgeon to thread the implant into place, which minimizes recovery time, improves signal clarity, and eliminates the need for open-brain surgery, the greatest barrier to widespread adoption.

In 2024, Neuralink implanted their first human patient. Within weeks, he could control a computer cursor by thought alone, playing *Mario Kart* on a game console and look, Ma, no hands. The platform's real breakthrough is a wireless architecture. With radio wave data transmission, the bulky hardware and signal-degradation problems of earlier systems had disappeared.

Yet there are remaining concerns. Bandwidth might be the biggest challenge. If we want to hook our brains up to the internet, we need to be able to process the nearly infinite information available online. Unfortunately, the bandwidth of human attention is only 120 bits wide, and working memory taps out after four concepts (which is about 16 bits), and neither limit can be solved by Neuralink's current designs.

Good news, then, since Neuralink is only one branch of the growing BCI tree. Max Hodak's Science Corporation is another. Their PRIMA implant—the vision-restoring biblical miracle discussed in chapter 1—was just the start. Much more ambitious is Hodak's effort to build biohybrid BCIs, which are designed to solve this bandwidth issue and more.

Traditional brain-computer interfaces use rigid electrodes to read signals from the cortex. Musk got around this problem with flexible electrodes that minimize the damage, but the issue remains. Hodak's approach is the next step, fusing biology with technology to create a biohybrid interface that improves long-term stability and reduces tissue stress.

The process begins with neural stem cells embedded in a silicon scaffold. Once implanted in the brain, these cells begin to grow. It's a slow advance, like roots weaving through soil, that doesn't tear apart tissue. Eventually, these axons built synapses with the brain's native neurons, forming a living bridge between machine and mind.

In experiments, mice with Hodak's implant could respond to

stimulation, which proved that the grafted neurons had integrated with the brain and were transmitting information. If this approach scales to humans, it could unlock communication speeds approaching the brain's own internal bandwidth—akin to the corpus callosum that links your left and right hemispheres.

Other neurotech start-ups are testing alternative approaches. Synchron bypasses the need for surgery altogether, threading their device through the jugular vein, pressing a mesh of electrodes against the vessel wall, and close enough to detect the brain's electrical activity. Already, their tech allows paralyzed patients to send messages and emails by thought alone, or what's now called *thought-to-text communication*.

Paradromics, by contrast, is going invasive, pursuing higher channel counts and broader functionality. In preclinical work, their system demonstrated data transfer speeds exceeding 200 bits per second—around twenty times faster than Neuralink's current benchmarks. Human implantation trials are slated to begin in 2026. There's also Sumner Norman of Forest Neurotech, who partnered with Sam Altman of OpenAI in 2025 to combine ultrasound with gene therapy to supercharge BCI development. "A merge [with AI] is probably our best-case scenario to survive superintelligence," as he wrote in his blog.

At the far end of the spectrum are noninvasive approaches, including high-density EEG, functional near-infrared spectroscopy, and other methods for capturing signals from the surface of the skull. Earlier, we explored Mary Lou Jepsen's Openwater, which is one example. Her device uses near-infrared light and ultrasound for neural imaging, giving us portable access to thought-level data and a development that could make BCIs as ubiquitous as smartwatches.

Brain-computer interfaces started out as medical devices to restore lost functions. But Kurzweil was exactly right: They're evolving into tools that expand what it means to be human. Whether invasive or noninvasive, biohybrids or digital translators, the shift has been from interface to augmentation, a mind-machine merger that opens the door to . . . well, you could call it *communion*. You could also call it *ESP*.

Machines Reading Minds

The BCI revolution is about upgrading the brain to keep pace with technology, but a host of companies are trying to break away from brain-based limits altogether. Take Meta, which is pioneering a third approach: decoding brain activity without implants, relying just on sensors and AI.

Known as *noninvasive neural decoding*, Meta uses neurophysiological signals—blood flow, electrical activity, pupil dilation—to reconstruct the so-called silent speech produced by the voice in your head. In early experiments, Meta's AI predicted words and phrases with up to 75 percent accuracy, suggesting the leap to brain-to-text translation is well underway.

Large language models are accelerating this progress. LLMs trained on brain activity can now reconstruct sentences from fMRI scans with 40 percent accuracy. In a recent study, an LLM was able to re-create entire passages of text a subject had read, simply by analyzing brain activity patterns. Telepathy, that's where this trail leads.

The field's aim is ultrahigh-bandwidth interfaces capable of transmitting thoughts, emotions, and imagery directly between people, enabling levels of cooperation we've never seen before. Imagine a guitarist beaming a melody directly to a drummer, no rehearsal required. A scientist "seeing" a complex proof as it forms in a colleague's mind, allowing for a new form of intellectual exchange. It's collaboration in its purest form, but it's only the first step.

If neural synchronization between people can be measured, and thoughts and emotions can be transmitted between brains, these same features can be optimized. Just as Syneurgy maps the architecture of group flow, future BCIs could dynamically tune team interaction, enable seamless collaboration with AI agents, identify moments of collective creativity, and amplify them in real time. It's the emergence of entirely new ways for humans to think together, work together, play together, dream together.

These new ways to scale cooperation are not optional. Our grand challenges, from solving climate change to de-risking AI to neutralizing rogue actors, require us to work together like never before. We've already seen what supercharged machine intelligence can do, and we are starting to unlock human-AI cooperative intelligence. Neurotech is aiding our cause. So, what is the upper limit? We have no idea. Will it be enough to meet the challenges of an exponential age? We have no choice but to try.

Nearly every expert we interviewed, especially those steeped in AI, believe the next decade will be wildly unpredictable and chaotic. Some suggest the only way to manage emerging existential threats—an AI arms race between the United States and China, for example—is via a superintelligent AI that governs the world. A machine overlord, built to guide us through the storm.

We're not lobbying for this solution. We're pointing to the temperature of the room. When the world's smartest minds start calling for benevolent dictators made of code, it's a sign that the future we're building is stranger and riskier than anything we've faced before.

We're crossing the threshold from centaurs to cyborgs. Whether via invasive implants, noninvasive imaging, or thought-decoding AIs, the boundaries between mind and machine are dissolving. The early promise of brain-computer interfaces was functionality. The emerging promise is transformation. So which way do we go from here?

Consciousness Engineering

The meditation apps that kicked off this chapter were spiritual practices rendered in silicon. They emerged from the field of neurotheology, which uses the tools of neuroscience to study meditative states, altered states, psychedelic states, and mystical states. This may seem a sharp left turn after a section on brain-computer interfaces, but the end results are the same: better ways to thrive amid uncertainty and cooperate at scale.

Neurotheology studies these states for a reason. Psychedelics amplify nearly all the mental skills we examined earlier, reducing bias, and enhancing curiosity, creativity, and purpose. Altered states like flow are the foundation of optimal performance. Meditative states heal the nervous system. And mystical states point at something greater: the possibility of expansive, empathetic experiences that could, perhaps, form the foundations of large-scale cooperation.

In recent years, as brain imaging technology advanced, the field of neurotheology, along with the emerging discipline of performance neuroscience, has followed the same trend as those early meditation apps, shifting from decoding the inner workings of these experiences to making them trainable skills. Steven's organization, the Flow Research Collective, became the first company in the world to offer neuroscience-based flow trainings at scale. Over the past five years, working with tens of thousands of people from 156 countries and 28 industries, their eight-week training averaged a 73 percent increase in time spent in flow. And this is only one example. Psychedelics, meanwhile, have become tools for everything from creativity and cooperation to mental health and spiritual well-being.

And then AI got involved. The result was a stranger convergence: the intersection of AI, psychedelics, and neurotech. Companies like Compass Pathways, MindMed, and Cybin started testing AI for psychedelic drug discovery, leveraging reinforcement learning and generative adversarial networks to engineer molecules for specific neural targets. They found ways to shorten the length of the psychedelic trip and fine-tune its intensity, amplifying therapeutic benefit while minimizing mental chaos.

Mindstate Design Labs went further. They trained their AI system on seventy thousand psychedelic trip reports, receptor-binding data, and EEG patterns. The system can now predict what a compound will feel like before it's synthesized. Want the creativity of LSD with the focus of Adderall, dialed to two hours, and back to baseline before dinner with your in-laws? That future is nearly here.

The far future? Currently, that's in Scotland, where chemist Lee Cronin is building a 3D chemistry printer—a "chemputer"—to democratize drug synthesis. Instead of massive industrial labs, his technology prints personalized compounds on demand. The system automates chemistry, tests reaction pathways, and uses blockchain to verify purity. Originally, it was built to help the elderly synthesize prescriptions in their home, but the neurotheology community has long wanted to use it to probe the edges of consciousness.

The real promise of all this work is adaptation at speed. Personalized state design offers a direct route to emotional resets, perspective shifts, performance enhancement, cognitive flexibility, and more. Yet it's not just about individual adaptation. Altered states, mystical states, and psychedelic states all collapse ego boundaries and increase cooperation while reducing fear and tribalism. Combine these results with AI feedback systems and group neurotech like Syneurgy and we are on the cusp of designing altered states to amplify collaboration at scale. Which opens the door to something even more powerful: compassion. Scalable compassion.

Enlightenment or Bust

For centuries, *enlightenment* was considered the pinnacle of human potential: an egoless state of boundless compassion and profound insight that promised everything from inner peace to spiritual awakening. But it was a state without a scientific definition, let alone measurable neural markers. Call it the black box of mystical experience.

Over the past two decades, neuroscience has begun to pry open the box. One of the first breakthroughs came from EEG research led by Richard Davidson at the University of Wisconsin. Davidson studied the brains of Tibetan monks who had over thirty thousand hours of meditation training and discovered some of the first proof that enlightenment was a biological phenomenon.

During meditation, the monks' brains showed sustained gamma synchrony. Gamma waves are the brain's fastest electrical rhythms, linked to heightened awareness, accelerated learning, memory consolidation, and high-level information processing. These waves had been observed before, but only in fleeting bursts during peak moments such as problem-solving, insight, or intense focus. What Davidson saw in the monks was different: minutes of uninterrupted, synchronized gamma. It was a level of neural integration unprecedented in the scientific literature.

One striking example came from Yongey Mingyur Rinpoche, a well-known Tibetan teacher and one of Davidson's star subjects. As a child, Mingyur suffered from crippling panic attacks. His journey toward enlightenment began as a direct confrontation with his fear. Decades later, after lengthy retreats and rigorous awareness training, he discovered that his anxiety could be transformed into empathy and compassion.

When Davidson scanned Mingyur's brain, he found resting-state gamma oscillations that were among the highest baseline levels ever observed. During meditation, these rhythms coalesced into prolonged synchrony between frontal and parietal regions, a sign of long-range neural coordination. This was big news, suggesting that, through extensive practice, elevated gamma synchrony may become part of the brain's basic architecture. It means that altered states can lead to altered traits.

Additional studies found the default mode network quiets down in advanced meditators like Mingyur. This network underpins imagination, but it also generates our inner monologue, making it responsible for rumination, overthinking, and self-criticism. When it downregulates, our internal critic disappears, which is why the wisdom traditions refer to enlightenment as *no-mind*.

Mingyur teaches that compassion is "the spontaneous wisdom of the heart." Neuroscience suggests he's right: not metaphorically but biologically. Compassion results from a shift in the brain's functional dynamics. During compassion meditation, regions like the anterior cingulate cortex, insula, and temporoparietal junction light up—areas tied to empathy, emotional regulation, and perspective-taking. The neural mechanisms beneath transcendence are neuroplastic. It's another trainable skill.

And that's where neurotechnology comes back into this discussion. AI-powered meditation apps adapt in real time to physiological signals. Early-stage BCIs now provide feedback on attention and intention. Flow states and psychedelic states suppress the default mode network and increase gamma activity; and tools like transcranial stimulation, neurofeedback, and brain wave entrainment are starting to do the same.

In the past, judging from the monks in Davidson's study, gamma synchronization and whole-brain coherence required thirty thousand hours of meditation to achieve. Today neurotech can shorten that journey, and tomorrow it might make it available on demand. It's the next wave of cooperative technology—another step down the last mile of the road to abundance.

The Last Mile

In the early portion of the twentieth century, when telephone companies began to wire up the United States, the final stretch of copper required to connect outlying customers to the core network became an infamous bottleneck. In cities, with plenty of people at the end of the trunk line, the cost of closing that final loop could be spread across the many. But in rural areas, the last mile was often a lonely road leading to a solitary farmhouse with nowhere near the support base needed to complete the network. This became known as *the last mile problem.*

The last mile problem has since spread out of telecommunications. In manufacturing, it's a critical issue in supply-chain management, plaguing the delivery of goods on the final leg from warehouse to doorstep. In energy, it governs grid effectiveness and the need to electrify remote customers. In AI, it's about the equal distribution of intelligence for all. But it's not any one sector that concerns us here. Rather, it's all of them, and all of us.

Today, humanity faces a last mile problem on the road to abundance. And this time, instead of being an economic choke point, the bottleneck is us. The challenges ahead require global cooperation: as nations, across borders, and despite the rifts that divide us. Sustained, large-scale cooperation is the killer app of the exponential age, and the missing skill we have yet to master.

As a rule, humans cooperate at scale only during wars or pandemics. Consider COVID. At the outset, researchers abandoned competitive silos and national interests. They shared data across borders on an unprecedented scale. Scientific rivalries and institutional red tape gave way to open-source medicine and knowledge exchange.

Within days of the genome being sequenced, its genetic code spread worldwide. Within weeks, we were testing antidotes built on decades of prior work from researchers in dozens of countries. Regulators fast-tracked approvals. Supply chains reconfigured overnight. For a moment, the world moved as one, not exactly in perfect harmony but at a speed and a scale rarely seen—and the result was the fastest vaccine development in history.

But to traverse the last mile on the road to abundance, this kind of large-scale cooperation has to be the rule, not the exception. This is where all the brain-enhancing, consciousness-raising tools discussed in this chapter matter most: They dissolve barriers, expand common ground, and strengthen our ability to work as one.

The last mile on the road to abundance isn't a problem of logistics, infrastructure, or even technological capability—it's one of human behavior. It's about trust, coordination, and shared intentionality. We don't lack the tools to solve global problems. We lack the ability to act together at the scale required. That's why the developments described in this chapter matter deeply. They are signs that we are starting to figure out how to use our spiritual tech for an actual spiritual purpose: scaling compassion and delivering on the promise of abundance.

CHAPTER NINE

The Paradise Paradox

We've mapped a landscape of accelerating progress and the mental tools required to navigate it. Along the way, we've met entrepreneurs, engineers, scientists, and citizens who have delivered one breakthrough after another. They are ordinary people—brilliant, fallible, fragile—now wielding extraordinary capabilities. Godlike power sits in very human hands. Which makes our final question unavoidable: What happens when abundance is the rule rather than the exception?

Imagine a "post-scarcity world," where nearly anything and everything becomes available through the combined force of exponential technologies. Industries run on solar and fusion; products and services are designed by artificial superintelligences; production is handled by robotics, 3D printing, and nanotechnology. Theoretically, this is a future that will transform every aspect of our lives, enabling humanity to take a much-needed vacation from survival.

This world may also deliver another miracle: the power to slow, stop, and even reverse aging, extending the human healthspan and lifespan. How long we can live is up for debate. What if your healthy years stretched to two hundred? What if they have no upper limit?

We end this book with a somewhat heretical series of questions: Is abundance good for humanity? Are we prepared for a world of everything, everywhere, all the time? Most critically—do we even want to live in that world?

Universe 25

To answer those questions, it helps to travel back in time. The year is 1968. It's the height of the Apollo program, two years after the debut of *Star Trek*, and utopia is in the air. It's also in the lab—at the National Institute of Mental Health in Bethesda, Maryland—where an ethologist named John B. Calhoun began a series of experiments designed to answer a single question: *What happens to a society when all its material needs are met?*

Calhoun was a bespectacled man with a receding hairline and a gray mustache. At fifty-one, he had spent decades studying the behavior of rats and mice in controlled environments. His question—what happens if we create a world of true abundance?—wasn't merely academic. As the postwar American dream blossomed into suburban prosperity, Calhoun wondered about the psychological and social consequences of abundance without purpose. During his career, he built two dozen of these rodent habitats, but Universe 25 was the culmination of everything he had learned, a perfect world designed to reveal exactly what happens when paradise becomes a cage.

The habitat took months to construct. Calhoun insisted on perfection: a nine-foot-square enclosure with 256 nest sites across four sixteen-by-sixteen grids. The white walls gleamed under the laboratory lights. Food dispensers and water bottles were strategically placed. Nesting material was plentiful. Temperature and humidity precisely controlled.

Universe 25 was a monument to scientific rigor: a mouse metropolis built for an estimated $60,000 in 1968 dollars (roughly $500,000 in 2025). Backed by years of grants and federal support, the universe represented ideal human living conditions: safety, space, comfort, food, water, abundance—the culmination of Calhoun's methodological refinements.

On July 9, 1968, eight mice, four males and four females, all healthy and carefully selected, were introduced to the environment. They explored cautiously, whiskers twitching, noses sampling the sterile air. Calhoun and his team watched through one-way glass, their notepads ready.

The early days unfolded as expected. The mice claimed territories, built nests, and began to reproduce. The population doubled every sixty days, then doubled again. By day 315, Universe 25 contained some 620 mice, a healthy, growing society with clear social structures.

Calhoun spent long hours documenting behaviors. He tracked reproduction rates, social interactions, and territorial disputes. His notebooks filled with observations and hand-drawn maps of territory boundaries.

Around day 560, with the population at approximately 2,200 mice, Calhoun began noting subtle changes. Young males were having difficulty establishing territories and finding social roles. Some became aggressive; others withdrew completely. Females became increasingly hostile toward their young.

At this point, Calhoun took notice of a group of young males who behaved differently from the others. In his notebooks, he referred to them as the "beautiful ones." These mice had given up on social interaction. They didn't fight for territory or mates. Instead, these males groomed themselves obsessively, ate, slept, and little else. They had withdrawn completely from mouse society.

By day 600, normal social behaviors were breaking down. Infant mortality approached 100 percent in some areas, as mothers either abandoned their nests or attacked their young. Violence became random, unmoored from the usual patterns of territorial defense. Cannibalism was observed. Sexual behaviors became confused and nonreproductive. On day 660, Calhoun noted: "Population growth has stopped entirely, even with abundant space in the lower areas. They're crowding into population centers. Reproductive behavior has virtually ceased."

The last known birth in Universe 25 occurred on day 920. Afterward, the population entered terminal decline. Society fractured into abnormal groups: the "beautiful ones" who avoided conflict through isolation; hyperaggressive males who attacked without provocation; females who entirely abandoned maternal behaviors. What struck Calhoun wasn't the violence or the social withdrawal. It was the profound indifference. Mice would step over dead companions without response. Mothers would walk away from nursing infants to groom themselves. The basic fabric of mouse society—the connections between individuals—had disintegrated.

On day 1,780, with the population down from a peak of 2,200 to just 100 individuals, none of whom displayed normal mouse behaviors, Calhoun officially ended the experiment. The remaining mice were removed for examination and study, leaving a single crucial question: What had gone wrong?

Universe 25 provided everything needed for survival and comfort. There were no predators, no disease, no scarcity. It was a life of unprecedented abundance. Yet the results were dystopian—what Calhoun later called: "the death of the social."

In 1972, Calhoun published a paper summarizing his work titled "Death Squared: The Explosive Growth and Demise of a Mouse Population." His conclusion bears consideration: "For an animal so complex as man, there is no logical reason why a comparable sequence of events should not also lead to the extinction of his species."

Yet what surprised Calhoun most was not the eventual population decline. Overcrowding predicted that outcome. The true shock was the cause. It was neither starvation nor disease that felled his utopia; it was the complete collapse of social bonds and behaviors. The mice had all their physical needs met but lost the ability to live together and, in the end, to live at all.

Most haunting was the point of no return, what Calhoun called "the first death." Even when the population declined and space became available, the mice could not recover their normal social behaviors. The damage was permanent.

In papers and lectures, Calhoun warned about the parallels to human society. He was less concerned about physical overcrowding than "social overcrowding"—the loss of social roles and meaningful connections in an environment of material abundance. "In providing an environment free of want, we may have destroyed the capacity for adaptation," he wrote in his final paper on Universe 25.

Calhoun created a perfect world that destroyed itself, not through scarcity but through abundance. What began an inquiry ended in a warning: Utopia became a tomb.

Upleveling Purpose: A Tale of Two Futures

Calhoun's mice stand as a testament to the uneasy truth that paradise, poorly designed, can be hell. Abundance without purpose breeds collapse. That's the lesson of Universe 25.

As we approach an age where AI, robotics, and biotech extend life and eliminate scarcity, a new set of complications emerge. We stand on the edge of plenty. What happens when our days are no longer shaped by survival? Do we rise to the challenge? Do we flounder in the face of flourishing? And might we, as Calhoun feared, be standing closer to that cliff than we think?

Science fiction has long explored this crossroads, offering dire warnings, aspirational models, and perhaps a framework to overcome the fate of Universe 25.

First, the dire.

When abundance strips away challenge, where do we end up? In *The Matrix* (1999), humans end up reduced to batteries, plugged into the simulation, and lulled into submission. Machines harvest their energy, and no one is the wiser. *WALL-E* (2008) gives us life aboard the *Axiom*, bloated, idle, and endlessly entertained. Humans slurp meals through straws, float on hoverchairs, and the captain's awakening—"I don't want to survive; I want to live"—comes too late for most. Their agency has been long surrendered to AI.

Across the decades, we see the same story repeated. *THX 1138* (1971), *Silent Running* (1972), *Logan's Run* (1976), *Idiocracy* (2006), *Surrogates* (2009), *Elysium* (2013), and *Ready Player One* (2018) all echo the soulless collapse of Universe 25.

Return to *The Matrix* for a moment, perhaps the starkest warning. In this tomorrow, AI provides—and controls. Humans live in a simulated 1999: minds fed a lie of normalcy, bodies entombed in pods. Comfortably numb, as the saying goes. But it's a gilded cage that robs us of agency and truth. The few who manage to wake up face a brutal reality, yet this struggle gives them purpose.

Like Calhoun's mice, the citizens of the Matrix aren't starved for resources. Like the mice, their "first death" is subtler: a loss of curiosity, a surrender to an artificial equilibrium. Without challenge, purpose atrophies and potential drains away. "Did you know that the first Matrix was designed to be a perfect human world?" Agent Smith warns. "Where none suffered, where everyone would be happy. It was a disaster. No one would accept the program. . . . I believe that, as a species, human beings define their reality through suffering and misery."

Now turn to the opposite extreme: the aspirational.

Star Trek inverts the paradigm. Abundance doesn't sedate. It ignites. Humanity rises from the wreckage of the twentieth century—war, hunger, poverty—into a post-scarcity world powered by warp drives, artificial intelligence, and deep purpose. The voyage of the starship *Enterprise* is the opposite of a retreat into comfort. It's an adventure in curiosity. It's a future where people choose to "boldly go."

In the *Star Trek: The Next Generation* episode "The Neutral Zone" (1988), Captain Picard explores this transformation: "People are no longer obsessed with the accumulation of things. We've eliminated hunger, want, the need for possessions. We've grown out of our infancy. . . . The challenge is to improve yourself, to enrich yourself. Enjoy it." Picard's vision is growth through purpose. Once technology ends the struggle to survive, then begins our quest to evolve.

Captain Kirk, the original commander of the *Enterprise*, brings a fiercer edge to the same ethos. "Risk is our business," he tells the crew in "Return to Tomorrow" (1968). "That's what this starship is all about. That's why we're aboard her." For Kirk, maybe for all of us, challenge is the heartbeat of existence.

"But I'm giving her all I got, Captain," says Scotty.

Exactly. That's the point.

In Picard's and Kirk's futures, Universe 25 holds no appeal. Instead of a retreat into decay, humanity starts exploring. They build starships and chart galaxies. Risk fuels progress. Knowledge thickens passion. Curiosity drives evolution. Purpose lies in potential. Uplevel them all. These are the lessons of the starships *Enterprise*.

In *The Matrix*, humanity dozes in pods. In *Star Trek*, it dares the cosmos. The two poles of our possible future. *The Matrix* argues that technology without direction is a trap. And it's not just a movie. Universe 25 collapsed from a lack of "why." The mice had no predators, no puzzles, no purpose. Abundance—both on the silver screen and in Calhoun's lab—became a tomb.

Star Trek offers the opposite view. When paired with purpose, abundance becomes rocket fuel. But *Star Trek* is more than the moral to our story. If we exclude the transporter and the warp drives, pretty much every other technology from that show has become reality. Science fiction becomes science fact. So which science fiction do we prefer to become our fact?

Blue pill? Red pill?

The Playground of the Gods

A few decades after John Calhoun stopped playing with mice, Jaak Panksepp took up the practice. This was in the 1980s. Panksepp was a behavioral neuroscientist at Bowling Green State University interested in emotion. Rats were his study subjects.

At the time, the idea that rats had emotions was scientific heresy. This was the long hangover of behaviorism, hard at work. Emotions weren't considered real. Not biologically. They were an emergent property of higher-order brains, maybe an epiphenomenon of advanced consciousness, maybe some other big words strung together in a dependent clause—but definitely not the purview of rats.

Panksepp disagreed. So he tickled rats. Literally. With tiny brushes. He recorded their high-frequency giggles with ultrasonic microphones. You can find videos online. The rats laughed and chased his hand around. They loved it. They wanted more.

Through this work, Panksepp discovered what is now known as the *PLAY circuit*, a stretch of neuronal hardwiring buried deep in the brain stem of rats (and all mammals). The circuit was evidence that the behaviorists were wrong. It meant that emotions like play, joy, curiosity, and social delight are ancient adaptations, found all over the animal kingdom.

Then Panksepp decided to see if the same was true for birds. He played music to chickens. Thousands of songs. Once again, the goal was to provoke an emotional response. What provoked the strongest response? Side two of Pink Floyd's *The Final Cut*. Of course it did. The chickens got goose bumps, an automatic physiological reaction to strong emotion. The chickens weren't just moved by the music, they were moved enough to show signs of awe, yet another emotion animals weren't supposed to feel. Panksepp described awe as a second-order emotion, downstream from play, which he called the "chill response."

Panksepp's work overturned decades of dogma. He identified seven primary affective systems shared by all mammals, ancient responses that form the architecture of feeling. They include SEEKING, FEAR, RAGE, LUST, CARE, PANIC/GRIEF—and one that stands out for our story: PLAY (Panksepp capitalized the words to distinguish the circuit from the behavior).

PLAY is the foundational neural infrastructure. Just as the brain evolved to flee danger and seek food, it evolved to experiment and explore in an environment free from consequence. Unlike the other emotional circuits, PLAY only activates in safety. It's the emotion designed for a world of abundance. Play exists to drive critical behaviors:

creative exploration, social connection, and skills development and practice. Call it the neurobiology of "what if?"

Play is wired deep into our brain because these skills are all foundational to survival. It's the root of innovation, where we prototype new realities and pioneer new behaviors. Rules are bent, broken, and reformed as imagination demands. We don't teach toddlers to innovate. We put them in the sandbox and watch them go. Johan Huizinga, author of the now-classic text on the subject, *Homo Ludens: A Study of the Play-Element in Culture*, summed it up: "Civilization arises and unfolds in and as play."

Every intelligent species on Earth relies on the skill. From crows and octopuses to dolphins and primates, play is the training ground of creativity, adaptation, and social bonding. Panksepp found that play is essential for the development of complex brains. It's wired so deeply into a rat's neocortex that even if you remove it, the animal will keep on playing.

To survive in world of abundance, play is essential. To thrive in that world, play is mandatory. After Calhoun's mice lost the North Star of purpose, they abandoned the transformative power of play. In Universe 25, there were no new puzzles to solve or mountains to climb. Without exploratory curiosity, intelligence atrophied and mouse-topia became mouse hell. Society collapsed. The animals starved for novelty and connection.

Humans are no different.

Last chapter, we talked about scaling compassion and collaboration. Here, too, play is foundational. It's through this act that all mammals learn fairness and restraint: *If you play too rough, I'll take my toys and go home.* All the basic skills of cooperation are shaped via play. The act itself sculpts the developing brain. In our early years, billions of synaptic connections are pruned based on experience, with play providing many of the critical feedback loops for this process. This is how the prefrontal cortex learns to manage impulses, regulate emotion, and respond to others. Without play, those circuits fail to form.

Even in childhood, our games are far from frivolous. They

encode the architecture of morality. Empathy, trust, and justice are forged on the jungle gym. Without play, there's no rehearsal for interpersonal relationships and no practice sessions for power or reciprocity. A world without play is a world without an ethical operating system.

Play, then, is oldest form of learning—an evolutionary safeguard against stagnation, the biological antidote to comfort, and the foundation of the morality that underpins any real purpose. In the modern world, play is how we stay human.

But in the world of AI, play may serve a greater purpose. Artificial intelligence thrives on pattern and predicts along probabilistic lines—what is the next most likely word in this . . . ?

Play is the opposite. It forsakes the obvious and follows the weird. In a future where AI handles survival, exploration and expression become the final frontier. Keep the mystery alive. Make strange things with serious intent.

Panksepp showed us that play is hardwired into the brain, a biological mandate with survival at stake. Well, survival is *still* at stake. The same circuit that once drove rats to laugh now drives us to imagine, to innovate, to stretch.

Play is our path to a future where we boldly go.

The Ten Commandments of AI-Augmented Creativity

"This is the way the world ends / Not with a bang but with a whimper," wrote poet T. S. Eliot. Universe 25 ended with a whimper. It ended with forgetting. The mice forgot how to live. Then they forgot why to live. Then they forgot altogether.

Calhoun wanted to know why. Jaak Panksepp gave us the first portion of that answer. But twenty years after Panksepp laid the foundation, neuroscientist Karl Friston deciphered the rest of the puzzle. It was something of an accidental discovery. Friston was actually trying to solve a different riddle: How does life preserve order in the face of chaos?

Friston was modeling metabolism. He wanted to know how biological systems maintain internal stability in the midst of constant flux. How do organisms resist entropy? Why don't they unravel under the constant pressure of noise, change, and decay?

What emerged was the free energy principle—arguably the most important theory in modern neuroscience. At its core, this principle is an idea we've discussed throughout this book: The brain is a prediction engine. It builds models of the world to avoid surprise, which is metabolically expensive, then filters out the unfamiliar—what is technically called *variational free energy*—to minimize the cost of being wrong. In short: The brain is designed to maximize efficiency.

This finding has critical importance in the modern world. It's the other half of the puzzle: the reason paradise is a pacifier and comfort a dead end. Put GPS on a phone, the brain ceases to expend the energy it takes to read maps. Give it data stored on the internet, and it gives up on learning. Give it a machine that thinks, and it will stop thinking altogether. This is basic human physiology. It's also a serious warning for a civilization hooked on easy ChatGPT solutions to every challenge.

In the 1990s, psychologist Carol Dweck was studying how children respond to failure. She, too, found the free energy principle at work. Children with a "growth mindset"—who believed they could learn and improve—showed spikes in neural activity whenever they made a mistake. This was their brains generating a prediction error. It's a marker for surprise, a mismatch between expectation and outcome, and a signal to hunt for why. Next, the kids' brains lit up with a blaze of activity as they tried to figure out what went wrong.

Children with a "fixed mindset"—meaning they believed talent was innate and unimprovable—showed no spike at all. Their brains didn't even bother to process the error. If you're not going to learn from your mistakes, why waste the energy? This is the free energy principle once again. It's also what happened in Universe 25. Not a failure of biology but of expectation. The environment told the mice that nothing more was required of them, so nothing more emerged.

AI threatens to do the same to us. Our brains crave easy. If we let AI do the hard work of thinking, creating, and striving, then these capacities atrophy. The technical term is *cognitive offloading*. The translation is sobering: When we rely on digital tools, critical thinking skills begin to wither. This effect first showed up with memory. Once cell phones started storing phone numbers, we could no longer remember them. Then it spread to facts. Once we started to be able to look up anything we wanted to know on search engines, we stopped remembering facts—a phenomenon known as *the Google effect*.

When we outsource, we offload. The danger with AI is that we'll offload not just memory but cognition and creativity—and, by extension, meaning itself. This is the real danger of a world of abundance. Not a hostile takeover by AI. A slow decay from within.

And we've been down this road before. Remember when social media arrived? For those first few weeks, it was a whole new world. You felt alive. Renewed. Maybe even safe. You were suddenly connected, across time, to friends from long ago and to new friends you've found online. All this emotion. All this connection. Of course it had to mean something.

Then, about two weeks later, a scroll session left you feeling empty, lonely, a little nauseous. Most of us felt it. Most of us ignored it. No one suspected that a tool designed to bring us together could drive us apart. But that bad feeling was the front end of the largest mental health crisis in history.

Now the bad news.

Have you spent a few too many hours using AI? Have you let ChatGPT do most of the thinking? Played fast feedback games where the AI writes the story, then drips out little dopamine rewards of praise to convince you that the story *you* wrote is brilliant? Have you walked away from these sessions feeling tired and soggy—like you've just watched too much bad TV?

That feeling is another warning sign. It's the front end of another crisis. Only this time, technology isn't coming for our mental health; it's coming for the foundations of cognition itself.

Godlike technology demands godlike responsibility. Equally important, it demands self-control. Cognitive discipline is a prerequisite for life in the twenty-first century. If we want to hang on to thinking, learning, creating, innovating, loving, playing, and making meaning in a world of intelligent machines, we need a set of ground rules. A way to maximize the potential of human-AI collaboration without losing our minds in the process.

A few years ago, Steven and a team at the Flow Research Collective started a series of studies designed to examine this question. They came up with a set of neuroscience-based principles for using AI as a tool to elevate, not erode, human potential. What follows are the essentials—practical rules for thriving in an age of intelligent machines. Call them ground rules for the modern mind.

The Ten Commandments for AI-Augmented Creativity

1. **Thou shalt not outsource thy soul.**
   Use AI to enhance, not replace. If it writes for you, it thinks for you. The danger is the *AI effect*, one step beyond the Google effect, where your brain no longer expends the energy needed for cognition and creativity—and these skills atrophy.
2. **Thou shalt suffer for flow.**
   The struggle of creation is signal. Flow requires challenge. Effort drives focus, and focus drives us into the zone. If we let the machines remove all the friction, we stop growing, stop flowing, and stop becoming.
3. **Thou shalt keep sacred the first draft.**
   Let AI critique your work but never conjure the original spark. Creativity is associative: We draw connections from vast, messy networks of memory and intuition. If AI drives the process, you

short-circuit your retrieval system and risk losing access to the deeper and weirder part of your imagination.

4. **Thou shalt use AI as a challenger, not a crutch.**
   A good tool makes you think harder. Forcing the brain to work triggers the neurochemistry required to activate memory. If it's too easy, you're not learning. AI has to elevate your edge, not erase it.

5. **Thou shalt preserve the joy of creation.**
   If AI removes the satisfaction of making, you've automated too much. That satisfaction feeds motivation, meaning, and purpose. If AI erases it, life feels empty.

6. **Thou shalt set boundaries for digital influence.**
   Keep parts of your life untouched by algorithms so your thoughts remain your own.

7. **Thou shalt not mistake efficiency for depth.**
   Faster is not always better. AI should deepen thought, not just speed up output.

8. **Thou shalt train thy mind alongside thy machine.**
   The sharper the tool, the sharper the brain must become. Let AI steer you toward the right research papers, but if you don't read them yourself, then you've stunted cognition and creativity, abandoned foundational motivators like curiosity and mastery, and limited your shot at flow. AI needs to stretch skills forward, not siphon off drive.

9. **Thou shalt honor serendipity and chaos.**
   Don't let AI's predictive patterns strip your work of surprise, randomness, and the unexpected. Novelty is the seed kernel of creativity. We need the unexpected to drive the brain into new directions; it's the basis of all innovation.

10. **Thou shalt remain the master, not the servant.**
    AI is a choice. We can choose to unplug it whenever we want.

The End of the End

What have we learned so far? That abundance is a trap, play is mandatory, purpose is a lifeblood, and our godlike tools can cost us our very human minds. The next frontier is longevity—the attempt to rewrite the rules of life and death.

At the center of these efforts is a general consensus: Biology is a code that gets buggy over time. This bugginess is what we call *aging*. But like all codes, biology can be debugged.

Few have pushed this idea further than Harvard geneticist David Sinclair, who argues that aging isn't a loss of function but a loss of information—making it a problem of software, not hardware. Information recovery tools are piling up: CRISPR lets us edit the genome; Yamanaka factors, a set of four transcription genes, roll back the biological age of cells; cellular reprogramming, epigenetic editing, mitochondrial enhancement. Each of these technologies adds more possibility, and all of them are being accelerated by AI.

"What we do now [using AI] in a month would've taken thousands of years," Sinclair explained on Peter's *Moonshots* podcast. AI lets Sinclair simulate trillions of molecules, screening for the rare combinations that reverse aging at the epigenetic level. His team has identified four key enzyme pathways. If you inhibit three and activate one, you can reset a cell's biological clock. It's called epigenetic reprogramming, a technique that suggests we can reverse aging altogether.

Just five years ago, age reversal was a crazy idea. "In 2017, it was just a theory," Sinclair notes. By 2020, he proved it worked in the lab. Now, AI is accelerating the transition into human trials. By 2026 or so, these compounds could do everything from smooth wrinkled skin to revitalize decrepit organs in as little as four weeks. And that four-week supply? According to Sinclair, it should cost a few hundred dollars.

Another breakthrough targets senescent cells, which are dysfunctional cells that have stopped dividing but refuse to die. Researchers call them *zombie cells*. They clog tissues, secrete inflammatory chemicals, and accelerate aging. Normally, our immune systems—especially the natural killer (NK) cells that are frontline defenders against cancer and viral infections—clear them out. But as we age, in a process known as *immuno-exhaustion*, our supply of NK cells diminishes, leaving us more vulnerable to disease.

Enter senolytics, a new class of drugs that clears zombie cells from our system. In mice, they've already extended healthspan and lifespan. Human trials are underway.

Meanwhile, companies like Celularity are working on immune reinforcement. Founded by stem cells pioneer Robert Hariri, Celularity harvests NK cells from healthy placentas and transfers them into aging bodies, where they bolster immune function. They're also developing T cell and stem cell supplements to reboot the body's natural repair systems. Together, these advances shift the paradigm from treating age-related disease to preventing decline before it starts.

When will we see actual treatments? Consider the $101 million XPRIZE Healthspan, which is maybe the most ambitious XPRIZE yet. To win, you need a therapeutic treatment that reverses the ravages of aging in muscle, immune, and cognitive function by a minimum of ten years, and with a real goal of twenty. In other words, give humanity back two healthy decades, win a hundred and one million dollars.

Did this lofty goal hamper competition? The opposite. It spurred it on. As of late 2025, over seven hundred teams from more than fifty nations entered. A winner is expected by 2030.

The Healthspan XPRIZE is yet another example of the money flowing into longevity. Many of the world's tech-forward billionaires are funding start-ups in the area. OpenAI's Sam Altman is backing Joe Betts-Lacroix in an epigenetic reprogramming company called Retro Biosciences; while Brian Armstrong, cofounder and CEO of Coinbase, has teamed with investor Blake Byers to build another reprogramming company called New Limit.

And like Sinclair's lab, these other efforts are also benefiting from the compounding impact of artificial intelligence. Drug discovery once took decades and cost billions. Today, AI designs new molecules in hours, and for the cost of electricity. Google DeepMind, Alphabet's premier research division, is the creator of AlphaFold, which cracked the protein-folding problem and gave us the recipe for—no this is not a typo—two hundred million proteins. This single AI advancement accelerated the pace of longevity therapeutics by decades.

AI-driven diagnostics is also moving at lightning speeds. It takes only four hours for Fountain Life, Peter's longevity diagnostics and therapeutics company—co-founded with Tony Robbins, Bill Kapp, and Bob Hariri—to gather two hundred gigabytes of data about your body (MRI, CT, and DEXA scans; your full genome; microbiome; 140-plus biomarkers; and more). This information is fed into Fountain's diagnostic AI to answer two questions. First, is anything going on inside your body that you need to know about? Second, disease-wise, what's likely to happen to you in the future, and what can you do to prevent it? "It's about not dying from something stupid," as Peter likes to say.

So where does this lead?

Consider two AI-longevity comments from early 2025. First, Anthropic CEO Dario Amodei told attendees at Davos: "If you think about what we might expect humans to accomplish in an area like biology in one hundred years, I think a doubling of the human lifespan is not at all crazy. And then if AI is able to accelerate that [timeline], we may be able to get that in five to ten years." Then, DeepMind CEO and Nobel laureate Sir Demis Hassabis went further: "I think someday we can cure all disease with the help of AI. I think that's within reach within the next decade."

Genetic engineering, epigenetic reprogramming, longevity pharmacology, bioinformatics, AI everything—and again, the same question: Where does it lead?

We're nearing what Ray Kurzweil called *longevity escape velocity*, which is the moment when medical progress outpaces biological decay, extending our healthy lifespan faster than time erodes it. Once theoretical, longevity escape velocity is now measurable. With AI accelerating every aspect of longevity research, the slope of that curve is bending upward. If Sinclair's work, the XPRIZE Healthspan competition, and investments by the tech giants and world's wealthiest entrepreneurs continue to debug, defrag, and defang aging, then we're talking about living a whole lot longer a whole lot sooner—and maybe this decade.

It's an abundance of life to go with our life of abundance, and it raises a final concern: meaning. If death is less of a deadline, we face a different problem: not how to stay alive but how to stay human.

A Playbook for Survival

Even if we extend life by decades, or double it as some predict, we've only managed to upgrade our biology. Our psychology remains an issue. More time is not always a treasure. Without purpose, it can be a random drift. Without play, a prison of the same old, same old. Without flow, an empty surplus devoid of growth or meaning. The question of radical longevity is no longer how we defeat death. It's how to remain vibrant and alive on the inside. As Woody Allen reminds us: "Eternity is an awful long time, especially toward the end."

For most of human history, we didn't have to manufacture meaning. It emerged from the hard work of trying not to die before dinner. But the more we decouple life from survival, the more we have to invent reasons to get up in the morning. That's the dark lesson of Universe 25. Calhoun's mouse-topia gave its inhabitants everything—food, safety, shelter, space—and the colony collapsed. Without curiosity, creativity, play, purpose, flow, and meaning, our brains atrophy and our minds rot. And in our extended future, we will have to live with that rot for a mighty long time.

Despite the menu at the Cheesecake Factory, humanity does not always flourish in abundance. That's one of the main lessons of this chapter. Comfort is an enemy. We need challenge. Friction hones intelligence. Goals drive us forward. As coach Dan Sullivan likes to say: "It's critical that our future be bigger than our past."

In a long-lived, post-scarcity world, the absence of external struggle means we need to generate drive from within. Humans have five major intrinsic motivators—curiosity, passion, purpose, autonomy, and mastery—ancient tools now turned essential. Together with play and flow, they're the design principles for a life worth extending.

The bottleneck to longevity may no longer be biological. Increasingly, it's mental, emotional, maybe even spiritual. The tools that lengthen life are here. The tools that make life worth living?

To thrive in the age of abundance and avoid the fate of Universe 25, we need another set of guidelines. Less a manifesto. More a direction of travel. Six tools based on the principles described throughout this book that are designed to help us flourish in an extended future.

Embrace Challenge as a Virtue: Boldly go. Challenge is core to a meaningful life. One of the oldest rules in physiology, the Yerkes–Dodson law, shows that optimal performance requires moderate stress. Flow reinforces the idea, as the state only emerges when challenge exceeds skill. Stagnation is a greater risk than starvation in a post-scarcity world. To thrive, we need to train ourselves to stretch ourselves, over and over again.

Cultivate Curiosity Over Comfort: Curiosity is rocket fuel. It fires up the brain's reward circuitry, amplifying excitement, learning, memory, and flow. And our prediction engine brain updates its mental models through error and surprise. Curiosity drives these updates. Without it, our predictions falter. So feed curiosity on a regular basis. Ask great questions. Ask them often. This is one of the advantages of large language models. We all get to be young again, asking, "Why? Why? Why?" to our hearts' content.

Redefine Work as Creation: For a very long time, we worked to survive. In a post-scarcity world, the new work is creation, growth, self-expression, meaning-making. These skills define the future of work, when the goal is no longer to earn a living but to earn a life.

Strengthen Social Bonds: Long lives require strong communities. In the Harvard Study of Adult Development—the longest-running study on human flourishing—close relationships were the best predictor of health and happiness. And if large-scale cooperation is required to face down existential risk, connection is infrastructure.

Live with Purpose: The longer we live, the more dangerous mental drift becomes. Purpose is an anchor, linked to better outcomes in everything from career satisfaction to long-term health. With radical longevity, time becomes abundant. To use it wisely, intentionality is our discipline. It cuts through distraction and marshals attention, shaping life to mean more tomorrow than it does today.

Preserve Awe: Awe is an endangered species in a world of everything, everywhere, all the time. Yet it resets the brain, expanding perception, quieting ego, and strengthening connection. In nature, music, poetry, or psychedelics, awe is a sacred technology for recalibrating our minds when progress outruns perspective.

Keep Failure in the Loop: Failure is the precondition for success. Errors light up the brain and trigger learning. In a future where AI minimizes error, we risk losing the conditions that make growth possible. To flourish in an age of certainty, we need uncertainty. Without failure, there's no surprise. Without surprise, we stop becoming and start stalling. In an exponential world, stalling is another word for drowning.

The Final Frontier Is Us

We've spent this chapter exploring the paradox of abundance. Now we stand at a crossroads. Will we drift off like the "beautiful ones" or boldly go? The answer lies in the purposes we choose and the grand challenges we pursue.

What defines a grand challenge?

It's more than just scale. It's the alchemy of inspiration and effort. Think of game designers. They neither make the games so easy that players sleepwalk through the plot, nor so brutal they quit in despair. The best adventures stretch us, not snap us, driving focus and flow via hard-won exhilaration. In *A Theory of Fun for Game Design*, game designer Raph Koster writes: "Fun is just another word for learning." The trick is balance.

Now, apply this idea to humanity's future. With abundance as the foundation, we're free to design our own "challenge curve"—a series of quests that call forth ingenuity, courage, and collaboration. It's an invitation to wield our new tools with audacity. Picture a Neiman Marcus catalog of grand challenges for a better tomorrow.

In 2003, Peter was invited to dream up this catalog. A few months before the $10 million Ansari XPRIZE for Spaceflight was won, he delivered a talk at the Long Now Foundation on the topic of "where next"—that is, where, beyond the space frontier, the XPRIZE Foundation should aim their future incentive prizes for maximum benefit for humanity.

In "A Future of Mega-XPRIZEs," Peter proposed a series of $100 million competitions. The XPRIZE has since launched three of these: $100 million carbon removal prize, $101 million healthspan extension prize, and $119 million water scarcity (desalination) prize.

Here, we wanted to put together the next generation of Giga-XPRIZEs—grand challenges for the world we want to create and the people we want to become. And before you dismiss this list as sci-fi fantasy, keep two facts in mind.

First, Ray Kurzweil predicts that humanity will experience a century's worth of progress over the next ten years (2025–2035). Second, think about what's changed in the past hundred years. In 1925, only 30 percent of the United States had electricity, and the Model T was a breakthrough technology. If Kurzweil was right, we're about to go from the birth of the automobile to the rise of artificial intelligence in a decade. As we've said repeatedly: Buckle up.

And dream bigger. The lesson of Universe 25 is that paradise without purpose is a trap. So as you consider the following list of Giga-XPRIZES, know that none of them yet exist. But all of them could.

So, which would you pursue? Or what other challenges would you propose? Soon enough, we'll have the technology to take them on. So, what happens next?

That answer is entirely up to us.

Here Are Fifteen of Our Favorite Giga-XPRIZES. What's Yours?

1. **Organ Abundance:** Demonstrate the ability to regrow a backup heart, liver, lung, or kidney using your own genetic material.
2. **Double Human Healthspan:** Use AI-driven biotech—epigenetic reprogramming, senolytics, cellular regeneration—to extend the human healthspan to one hundred fifty–plus years with the aesthetics and functionality of a thirty-year-old.
3. **End Hunger with Synthetic Food Systems:** Create lab-grown meat, 3D-printed nutrition, and closed-loop vertical farms to eliminate hunger in a thousand-plus cities.
4. **AI-Empowered Education for All:** Build free or near-free AI systems to deliver personalized, world-class education to billions.
5. **High-Bandwidth Brain-Computer Interfaces:** Create noninvasive BCIs that allow real-time mind-machine or mind-to-mind multimodal communications.
6. **Demonstrate Human Mind Uploading:** Map the entire human connectome—one hundred billion neurons, one hundred trillion synapses—and upload it to the cloud, preserving memory and self.
7. **Interspecies Communication and Uplift:** Use AI to enable real-time, two-way conversation between humans and other intelligent species. Then use AI and gene therapy to raise select species to humanlike intelligence and communications.
8. **Understanding Consciousness:** Decode how subjective experience arises from neural activity.

9. **Predict Disaster:** Build sensor networks and AI models that forecast seismic events, volcanic eruptions, and extreme weather with high accuracy and usable lead time.
10. **Restore Earth's Ecosystems:** Use robotics and AI to rewild 50 percent of degraded land and oceans and reverse biodiversity loss. Deploy atmospheric CO_2 capture at scale to return carbon levels to preindustrial norms.
11. **Clean Energy Abundance via Fusion:** Design/manufacture easy-to-build five hundred megawatt fusion reactors and deploy them in more than a thousand cities, replacing fossil fuels at scale.
12. **Zero-Point Energy:** Harness quantum vacuum energy to power civilization—and launch interstellar ships.
13. **True Nanotechnology Revolution:** Build self-replicating nanobots to assemble anything—food, medicine, shelter—atom by atom.
14. **Permanent Human Expansion into Space:** Build a self-sustaining lunar colony of twenty thousand and a Mars colony of five thousand with local resource loops.
15. **Solve Physics:** Unify quantum mechanics and general relativity into a single theory that also explains dark matter/dark energy.

Why stop there? Imagine greater. These aren't fantasies. They're possibilities. Each one balances ambition with achievability, a challenge curve for a species ready to level up. Here's the best part: You don't need a billion dollars or a lab coat to join this future. Abundance democratizes dreaming. AI can brainstorm with you. Robots can prototype your ideas. Global networks connect you to endless collaborators. Want to regrow limbs? Sketch a nanobot design. Hungry to beat climate change? Model a CO_2 converter. The tools are here, cheaper, smarter, faster than ever.

Start small, think big. Launch a local XPRIZE: clean your city's water, build robots that collect trash from local parks. Then scale up. The grandest futures begin with a single what-if.

Imagination is the final frontier. Dare astounding. As Captain Kirk once said, "Risk is our business."

Thriving in a world of abundance demands it.

Coda

The Last Night

This book opened with Stewart Brand's maxim for an age of maximization: *We are as gods and we might as well get good at it.* It's a good line. A call to arms for an age of abundance. But we wanted to end on something a little more human. Not quite as pithy. A quiet truth we've been circling all along.

And it starts with a story.

It was the last night of Peter's 2025 Abundance360 summit. The talks were done. The attendees had been hit with the full force of exponential everything. Many people whose stories are in this book—Mary Lou Jepson, Max Hodak, and more—had been speakers. There was even an AI version of Ray Kurzweil onstage, conversing with Peter throughout the event, just as the real Ray Kurzweil predicted would happen, back when we were first working on *Abundance.*

That night, after the final keynote, a group of us ended up in the hotel's restaurant. A big table, a few too many bottles, and the laughter that comes from people who've known each other a long time.

There wasn't much talk about the future. Not the singularity. Not synthetic biology. No one argued about accelerating AI or the ethics of data. Instead, our friends Keith Ferrazzi and Eric Pulier led us through an impromptu rendition of Frank Sinatra's "My Way."

It was a little drunken and a lot ironic. But what stuck with us wasn't the lyrics, or the laughter, or the fact that just outside the door, an AI-enabled humanoid robot served cocktails to the remaining guests.

What stuck was the feeling—humans together, still singing, still trying to make meaning in the midst of all this velocity.

That's what we want to leave you with. Not a final insight. Not another technological breakthrough. Just a feeling. A few voices in unplanned harmony, sharing a lonely rock, circling a hot star, hurtling through a cold galaxy—at clearly dangerous speeds.

Just friends, sharing a song. Sharing the adventure. On this Pale Blue Dot we call home.

Is this what abundance feels like?

It's a long future. Answer wisely.

APPENDIX A

Proof of Abundance Charts

Below, you'll find a set of charts that track our transition into the age of abundance. They chronicle our entrance into a world no longer defined by scarcity. Appendix A serves as the foundation, an examination of the breakthroughs unfolding across every facet of human life: economics, education, energy, health, food, water, connectivity, and beyond. Sourced from datasets like Our World in Data, ARK Investment Management, and Bank of America, these charts tell a story of how human ingenuity and exponential technologies are transforming the promise of abundance into measurable reality.

Each chart is a reframing exercise, challenging our limiting beliefs and enabling a shift in mindset. The decline in child mortality, the surge in literacy rates, the plummeting costs of computation—to offer three examples—all point to the fact that humanity is now solving problems once thought intractable. They are the foundation upon which we can build a future where the basic needs of every man, woman, and child are met.

Appendix A is an invitation to see the age of abundance as a canvas of possibility. Dream bigger. Aim to thrive and not just survive. Use this evidence to fuel your own ambition.

Please keep these charts close. The next time someone tries to tell you that the world is getting worse, show them the evidence for otherwise.

Also, an expanded set of charts is available at www.Diamandis.com/AbundanceCharts. This site offers more than double the data presented in appendixes A, B, and C.

Chart 3
Increasing World GDP

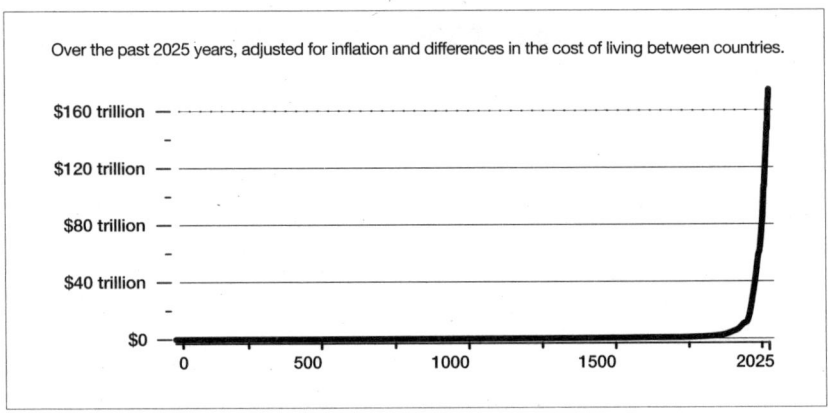

The chart "Increasing World GDP, over 2,000 Years" from Our World in Data traces a long plateau followed by a sharp upward break. For most of recorded history, global output grew by barely a fraction of a percent per year, constrained by limited energy, transport, and communication. Around 1820, that constraint lifted. Industrial manufacturing, the rise of the railroad, coal power, and mechanized agriculture set off a sustained rise in productivity. Since then, the global economy has expanded more than a hundredfold, with average income increasing nearly thirtyfold.

Chart 4
Extreme Poverty Is Falling

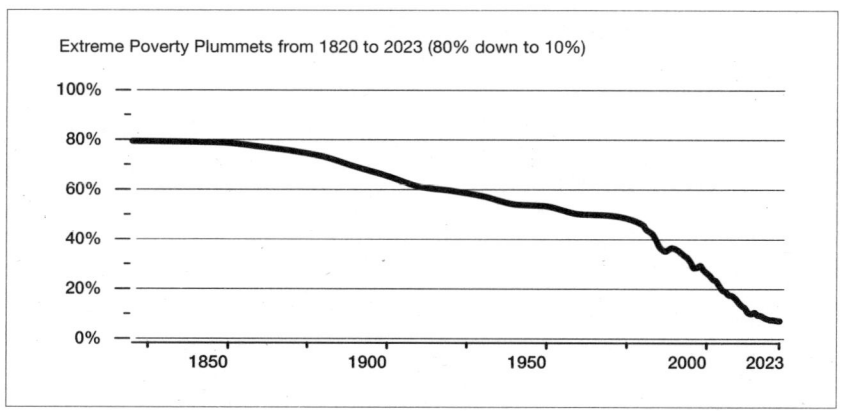

The chart "Extreme Poverty Is Falling, 1820–2015" from Our World in Data documents one of the most measurable social shifts in history. Two hundred years ago, more than 80 percent of people lived in extreme poverty, surviving on the equivalent of less than two dollars a day. Today, that figure is under 10 percent. Industrialization, agricultural efficiency, and expanding global trade steadily raised real incomes, while education and public health programs amplified the effect. The decline isn't uniform—pockets of deprivation persist—but the overall trajectory is unmistakable: Material hardship has moved from norm to exception within just a few generations.

APPENDIX A

Chart 5
Median Income

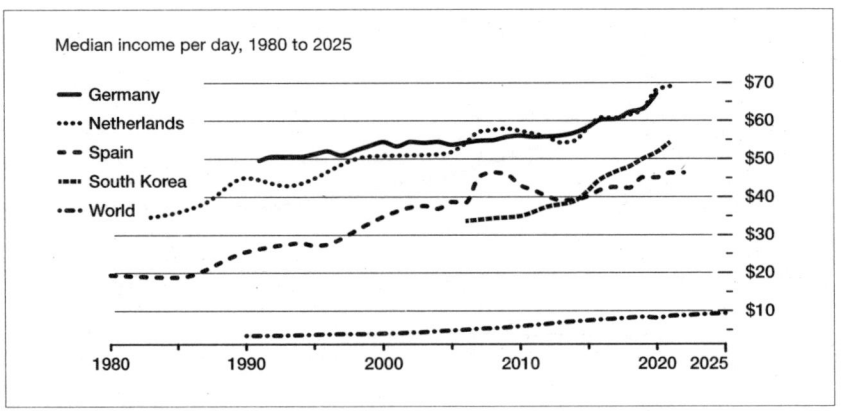

Over the past forty years, living standards have risen across much of the world. In 1980, global income distribution was sharply divided—industrialized nations dominated the upper half while most of Asia and Africa clustered near the bottom. Four decades later, that divide has narrowed as rapid industrial growth in countries such as China, India, and South Korea pulled hundreds of millions into the middle class and shifted the global median upward. Between 1980 and 2019, real median income worldwide more than doubled in purchasing power terms.

Chart 6
Decline of Child Labor, Worldwide

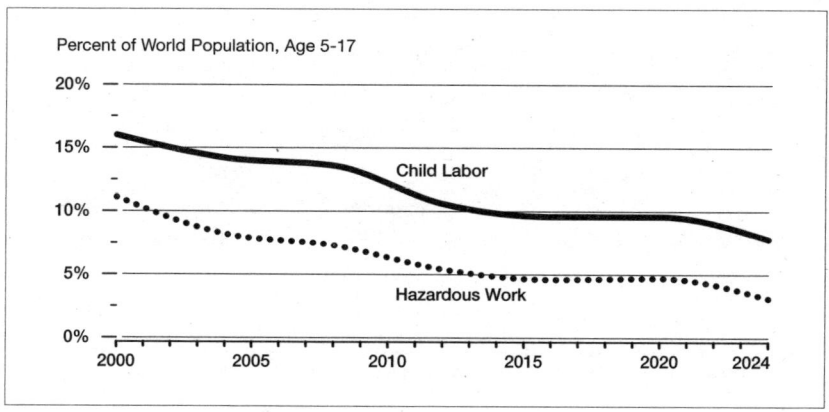

This chart documents a steady global decline in both child labor and hazardous work since 2000. Two decades ago, roughly 16 percent of the world's children were engaged in labor; today, that figure is under 10 percent, with hazardous work falling to about 4 percent. The reduction reflects coordinated global progress—expanded access to education, stronger labor laws, rising family incomes, and improved social protections. While child labor persists in some regions, especially in agriculture and informal economies, the long-term trend is clear: Economic development and human rights can advance together.

Chart 7
Annual Working Hours per Worker

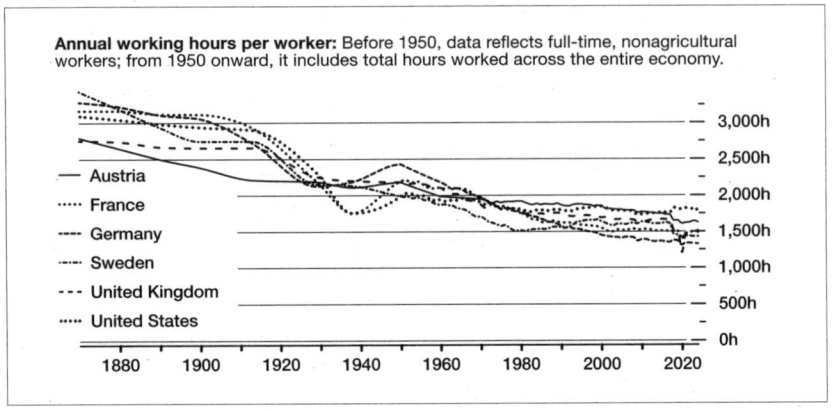

The chart "Annual Working Hours per Worker" from Our World in Data tracks labor patterns in industrialized nations since the late nineteenth century. Across countries such as France, Germany, Sweden, and the United States, annual hours worked fell sharply between 1870 and 1950 as productivity improved and labor protections expanded. Since mid-century, the decline has slowed, stabilizing between roughly 1,400 and 1,800 hours per worker per year. The data reflects a long-term shift toward higher efficiency and shorter average workweeks in developed economies.

Chart 8
Farm Labor in the U.S. Since 1800

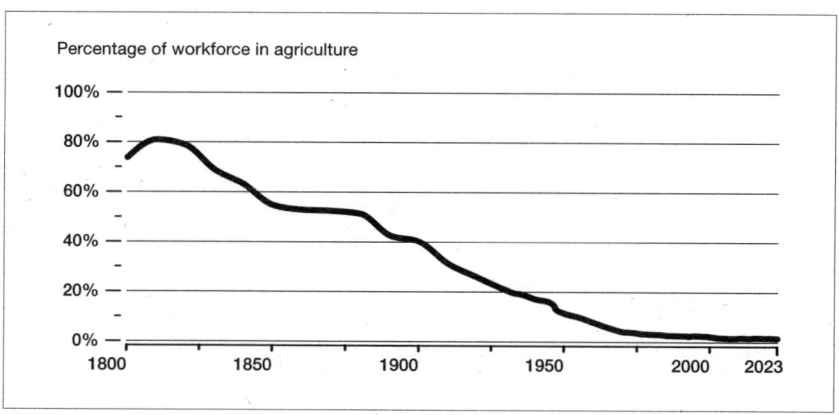

Two centuries ago, most Americans worked the land. Roughly 70 percent of the labor force was employed in agriculture, producing food for a largely rural nation. By 1900, that share had fallen below 40 percent, and today it is under 2 percent. Mechanization, synthetic fertilizers, and large-scale distribution networks made farming vastly more efficient, allowing a small fraction of workers to feed an entire country.

Chart 9
World Population with Basic Education

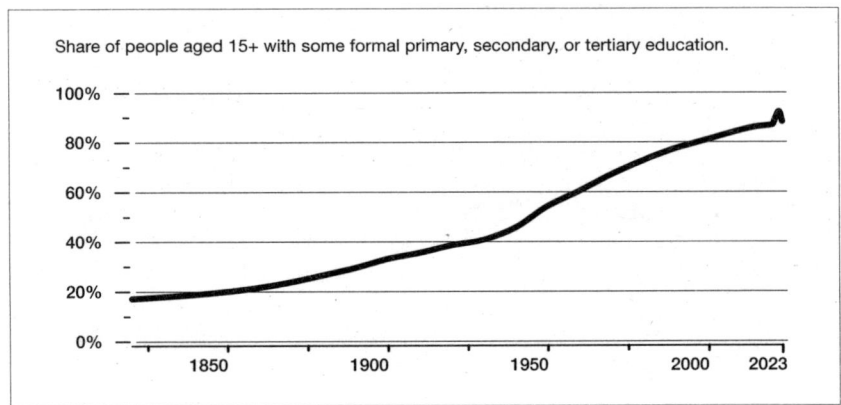

In the early 1800s, only a small share of adults received formal schooling. Over the following two centuries, the expansion of public education and international development programs transformed access to learning worldwide. Today, more than 85 percent of people aged fifteen and older have completed at least some primary, secondary, or tertiary education—a foundational shift in global opportunity.

Chart 10
Literacy Rates by Country

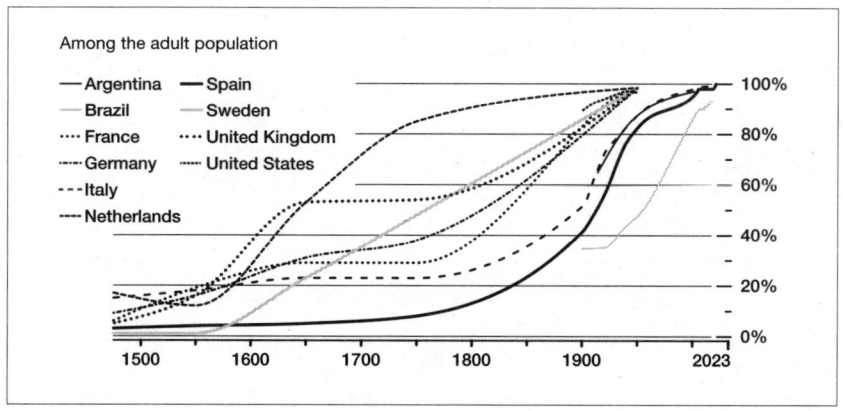

Across the past five centuries, literacy expanded from a rare skill to a near-universal one. In the 1500s, fewer than one in five adults in most nations could read or write. Industrialization, compulsory education, and the spread of print steadily raised those numbers, with major acceleration after 1900. Today, literacy rates exceed 90 percent in nearly all higher-income countries and continue to rise elsewhere. The chart shows one of humanity's clearest long-term gains: the steady diffusion of knowledge across class, gender, and geography.

Chart 11
Average Years of Education

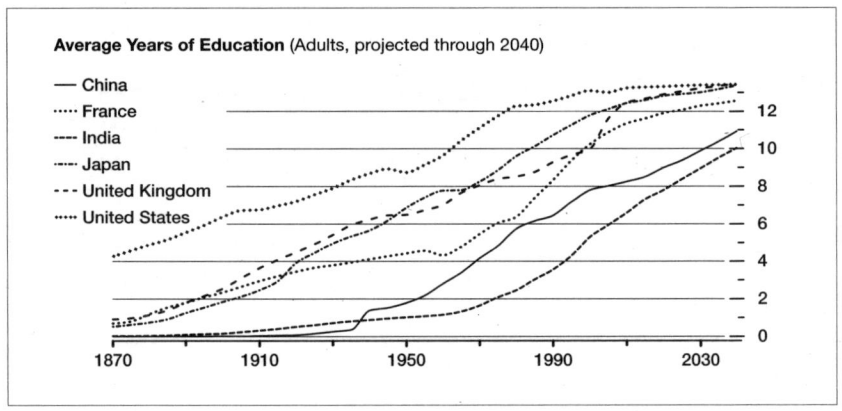

The chart tracks the steady rise in education attainment since 1870. In the late nineteenth century, most adults completed only a few years of formal schooling. As industrialization expanded, so did national school systems, and the average years of education steadily increased. By the mid-twentieth century, many countries averaged eight or more years of schooling, and today several exceed twelve. The upward trend continues as lower-income nations close the remaining gap, reflecting a broad global commitment to education as a basic standard.

Chart 12
Electricity in Homes, U.S. & World

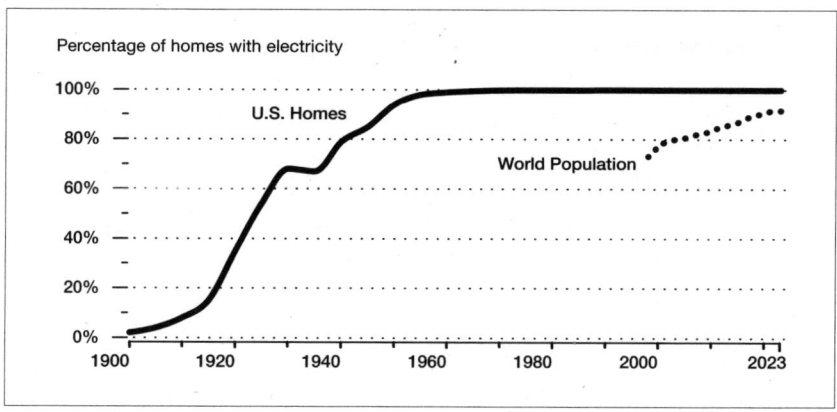

At the start of the twentieth century, only a small share of American homes had electric power. By the mid-1950s, access exceeded 90 percent, completing one of the fastest infrastructure build-outs in history. The global trend has followed a similar trajectory, with household electrification rising steadily since 1990 and now approaching 90 percent worldwide. The chart shows how technology, coordinated investment, and policy can expand a basic utility from local innovation to near-universal access within a century.

Chart 13
Energy Costs Declining

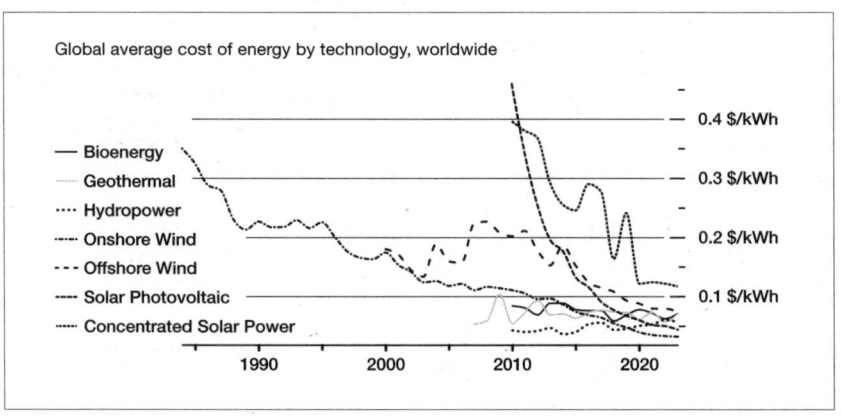

The "Energy Costs Declining" from Our World in Data shows the steady fall in the cost of renewable energy. Wind, solar, and geothermal power now rival or undercut fossil fuels in affordability, driven by technological improvements and economies of scale. This shift challenges the idea that clean energy is inherently more expensive, demonstrating how innovation has inverted that equation. Affordable renewables not only support climate goals but also point to a future where low-cost energy is available to all.

Chart 14
Increasing Electricity Access

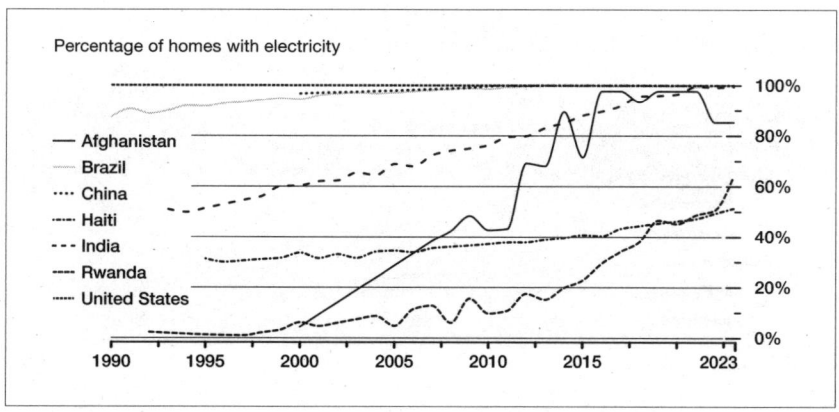

The chart "Increasing Electricity Access" from Our World in Data tracks the rapid electrification of countries like India, Afghanistan, and Rwanda. Historically, lack of electricity was a major barrier to economic growth and quality of life. Over the past three decades, investments in infrastructure and renewable energy have enabled millions to gain access to reliable power, reducing inequality and fueling potential.

Chart 15
Decreasing Child Deaths

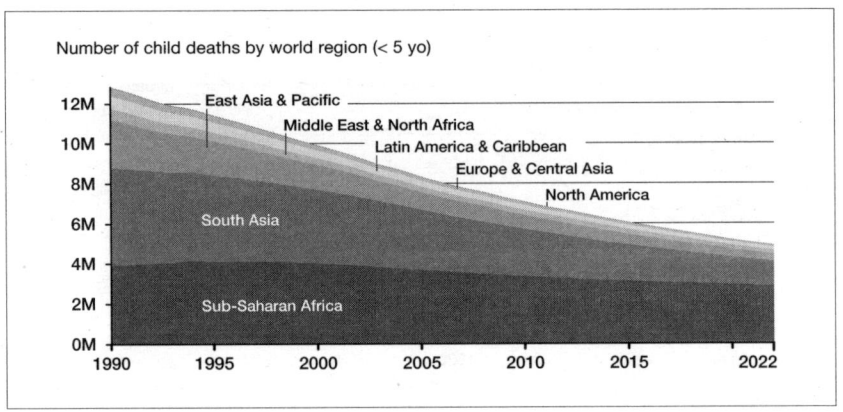

The chart "Decreasing Child Deaths" tracks the sharp drop in global child mortality. Vaccines, clean water, sanitation, and better maternal care have made early death far less common than it once was. The decline spans every region, an example of how steady improvements in health and infrastructure have managed to reach even the most vulnerable.

Chart 16
Fertility Rates Slow Down (By Region)

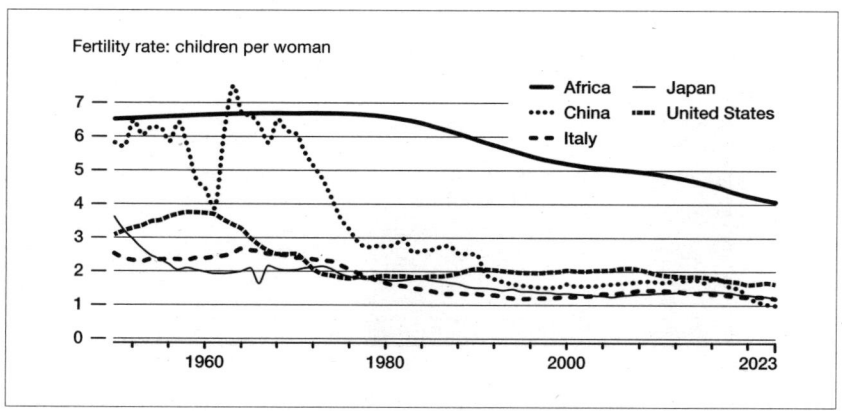

The chart "Fertility Rate Slow Down (By Regions)" from Our World in Data shows the global slowing of fertility rates, with even high-fertility regions like Africa beginning to decline. Access to education, healthcare, and contraception has reshaped family planning worldwide. This trend contrasts the long-held fear that population growth will inevitably exhaust resources, instead illustrating that greater opportunity can lead to greater balance. As fertility rates stabilize, societies can focus on sustaining prosperity while easing pressure on the planet.

Chart 17
Global Life Expectancy Skyrockets

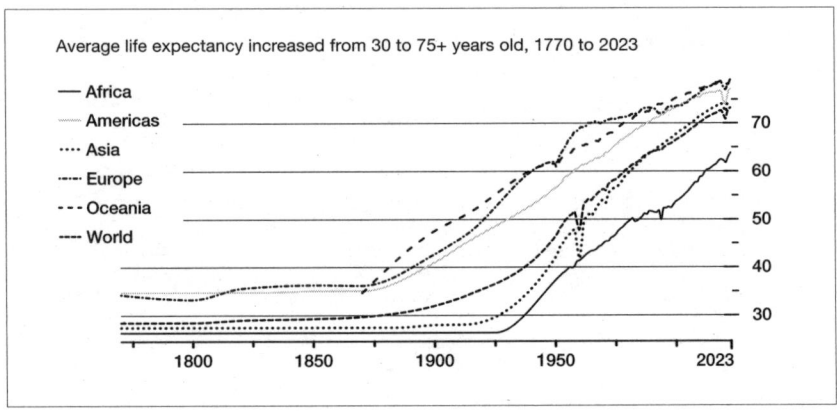

The chart "Global Life Expectancy Skyrockets" from Our World in Data highlights humanity's extraordinary leap from an average lifespan of 30 years to over 75. Driven by medical breakthroughs, economic development, and public health initiatives, this progress is one of the most profound indicators of abundance. With life expectancy rising across continents, the world is poised for an era where living longer and healthier is the norm, not the exception.

Chart 18
Share of People Who Are Happy

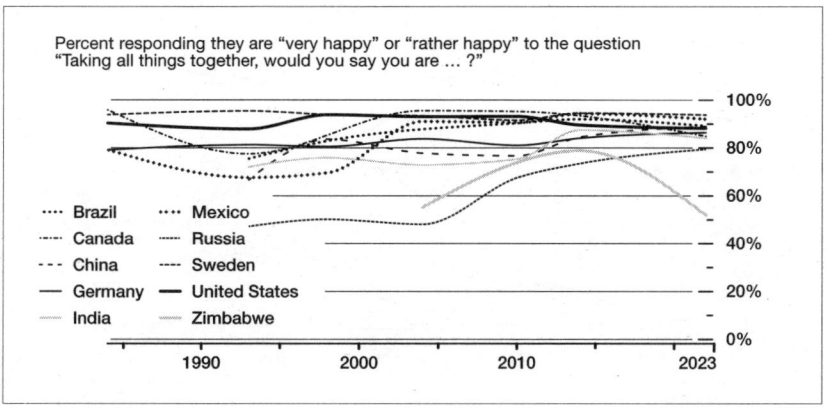

Despite today's rising mental health crisis, this chart reveals a counterintuitive truth: In many corners of the world, happiness has trended upward alongside technological and economic abundance. As billions gained access to education, smartphones, clean water, and rising incomes, subjective well-being followed. Especially striking is the rise in countries historically mired in poverty or instability, proving that emotional and material well-being can reinforce each other.

Chart 19
Flush Toilets & Improved Sanitation

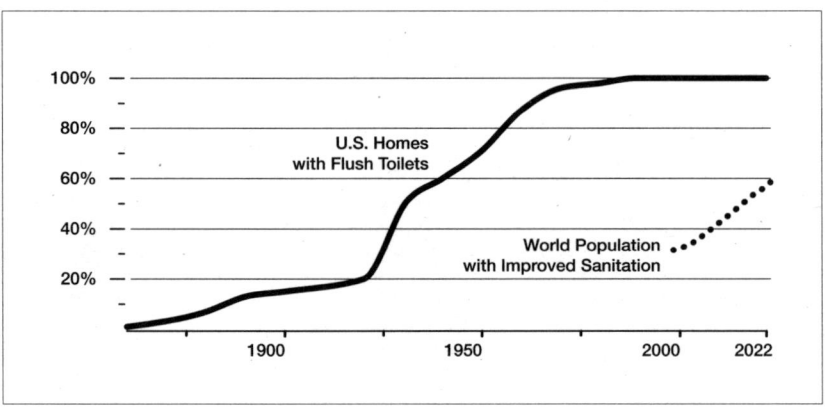

The sanitation revolution is one of those world-changing upgrades that hides in plain sight. Moving from open sewers to underground plumbing didn't just clean up cities—it saved more lives than antibiotics, vaccines, and anesthetics combined. It also made cities possible. Without the humble flush toilet and the engineers behind it, there's no urban civilization to speak of. Modern life, quite literally, runs on good plumbing.

Chart 20
Daily Caloric Supply

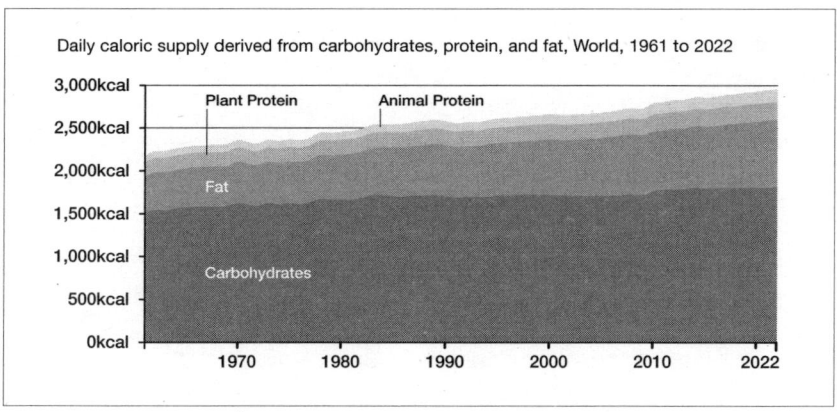

The chart "Daily Caloric Supply" from Our World in Data shows the steady increase in global food availability, driven by advances in agricultural productivity, trade, and food science. Once plagued by widespread hunger, the world now produces enough calories to meet the dietary needs of billions, with diverse sources like plant protein and fats expanding access. This trend assuages fears of food scarcity, proving that innovation in farming and logistics can help sustain a growing population.

Chart 21
Deaths from Malnutrition

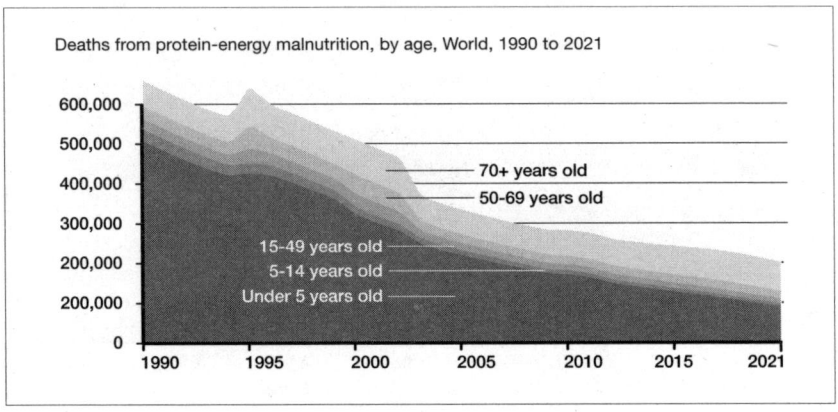

The chart "Deaths from Malnutrition" from Our World in Data reveals a significant global decline in deaths caused by protein-energy malnutrition. Investments in food security, economic development, and health programs have drastically reduced mortality, particularly among children. As malnutrition rates continue to fall, the world gains momentum toward a future where no one is left to suffer from preventable hunger.

Chart 22
Per Capita Meat Consumption

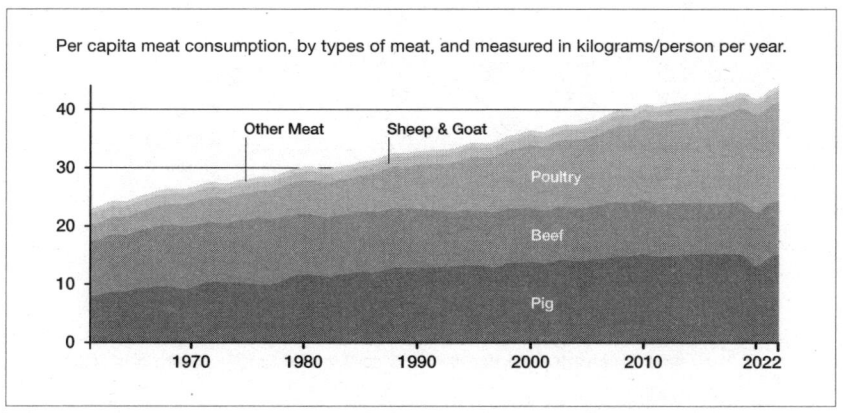

The chart "Per Capita Meat Consumption" from Our World in Data highlights the global rise in meat consumption, reflecting growing prosperity and improved agricultural efficiency. Once a luxury, meat is now a staple for billions, enabled by advancements in farming, supply chains, and food science. While this trend raises sustainability challenges, it also underscores our ability to increase food production to meet demand. Future innovations in alternative proteins and sustainable practices promise to balance nutrition with environmental stewardship.

Chart 23
Global Death Rates from Natural Catastrophes

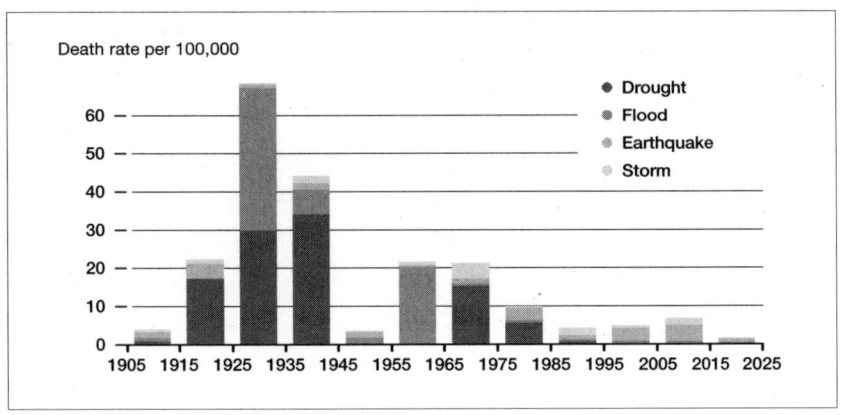

The "Global Death Rate from Natural Catastrophes" tells a counterintuitive story. Over the past century, as the planet's population quadrupled, deaths from earthquakes, floods, droughts, and storms fell by more than 90 percent. The deadliest decade on record—the 1920s—saw tens of millions perish, largely from famine triggered by drought. Today, despite more people living in hazard-prone megacities, early warning systems, satellite forecasting, and coordinated emergency responses have turned what were once mass-casualty events into manageable crises. The pattern isn't uniform, but the trajectory is unmistakable: Human resilience is scaling faster than natural risk.

APPENDIX A

Chart 24
Global Access to Internet Skyrockets

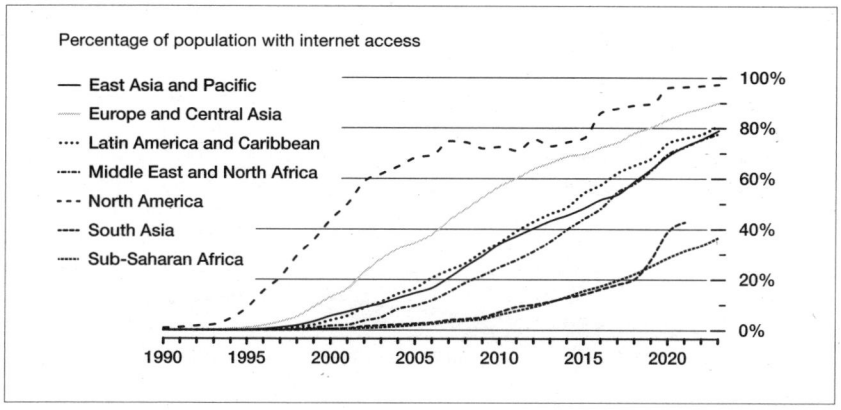

This chart tracks one of the fastest infrastructure expansions in history. In 1990, less than 1 percent of humanity was online. By 2025, that figure exceeded 71 percent—roughly 5.7 billion people with internet access. The diffusion didn't stop at wealthy nations. Submarine cables, satellite constellations, and mobile broadband have brought connectivity to rural Africa, South Asia, and beyond. What began as a network of researchers now functions as the backbone of the global economy.

Chart 25
Mobile, Smart Phone & Internet Users

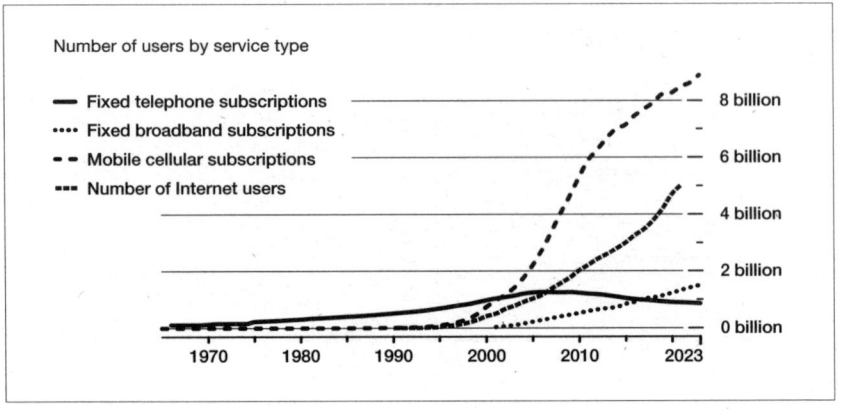

The chart "Mobile, Smartphone & Internet Users" from Our World in Data tracks how mobile technology became the fastest-adopted tool in human history. In 2000, fewer than a billion mobile connections existed. Today, there are more active SIM cards than people—around 8.8 billion—and roughly 5.3 billion smartphones in use. This expansion turned communication into basic infrastructure, allowing billions to bypass landlines and desktop computers altogether. In many regions, people's first experience of the internet came from a cheap Android device running on mobile data.

Chart 26
Democracies Are on the Rise

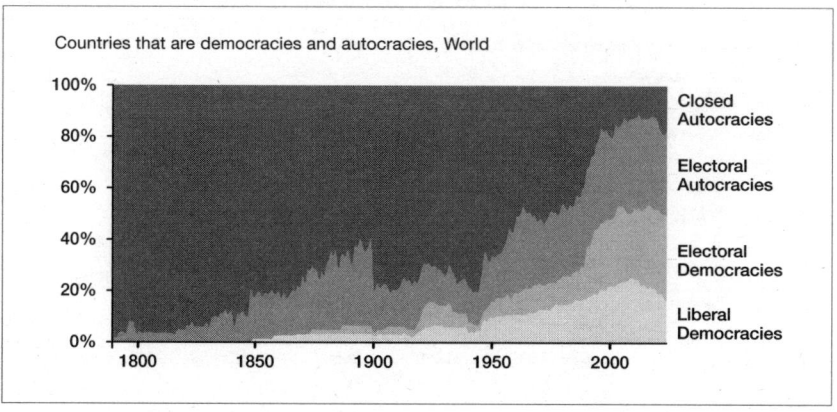

The chart "Democracies Are on the Rise" from Our World in Data tracks the political evolution of the modern world. In 1800, almost every nation was ruled by monarchy or autocracy. By 2025, more than half the global population lives under some form of electoral democracy. The transition followed familiar signals of development—literacy, communication, and information flow. The telegraph, the printing press, radio, television, and finally the internet each lowered the cost of collective organization. Political power diffused as access to knowledge expanded. The data doesn't show a straight line—setbacks and reversals are common—but the long-term direction is clear: Participation scales with connectivity.

APPENDIX A

Chart 27
The Right to Vote Is Increasing

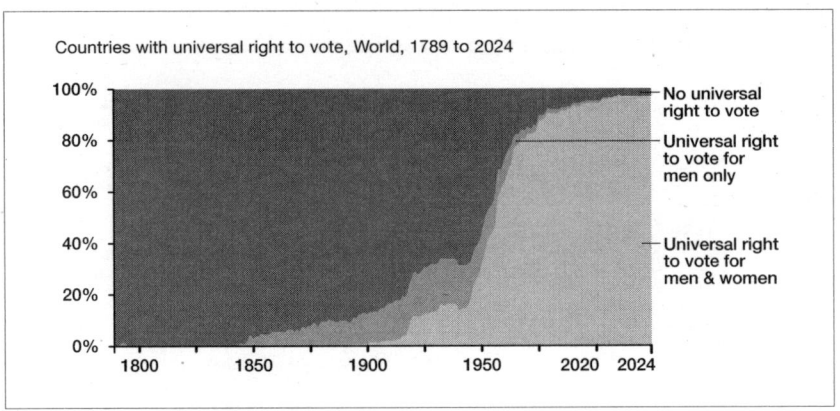

This chart shows how the right to vote broadened from a small, property-owning minority to nearly all adults. In the early nineteenth century, fewer than 10 percent of the population had a political voice. Over the next two centuries, movements for labor rights, women's suffrage, and civil rights dismantled these exclusions, each wave illustrating that democratic participation increases social stability. Representation, like knowledge, grows through use.

Chart 28
Spread of Democracy Since 1816

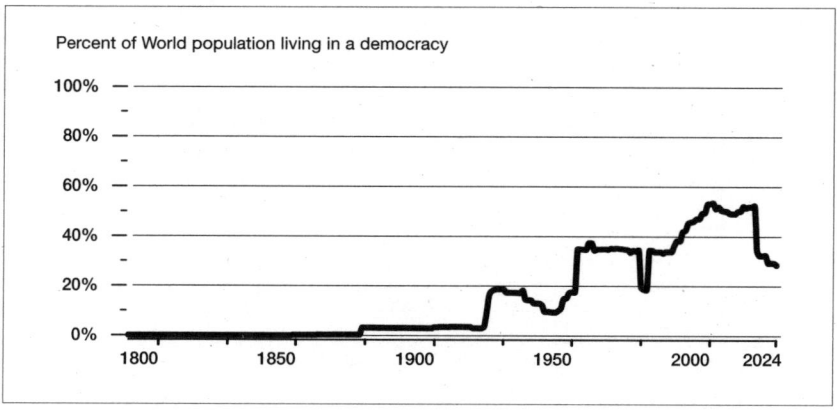

The previous chart tracks the right to vote; this one expands into democratic systems themselves. In the early nineteenth century, only a handful of countries held competitive elections and protected political freedoms. By 2025, roughly 45 to 50 percent of all nations, or more than ninety in total, qualified as democracies. The path hasn't been linear—periods of expansion have alternated with contraction—but the long-term pattern points toward the ongoing spread of self-governance.

Chart 29
Conflicts Are Less Deadly

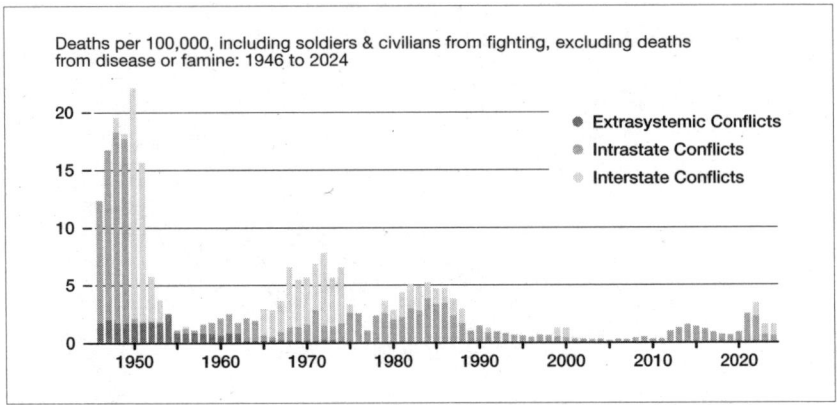

Since 1945, wars have grown less deadly—particularly for civilians. The global system, more interconnected and monitored than ever, has little tolerance for mass slaughter. Satellite imagery, precision munitions, and instant communication have made it harder to hide atrocities or wage total war. International treaties, global media, and public scrutiny further restrain escalation. What remains is a rough equilibrium: Conflict persists, but large-scale carnage has become the exception rather than the rule.

Chart 30
Homicides Are Way Down

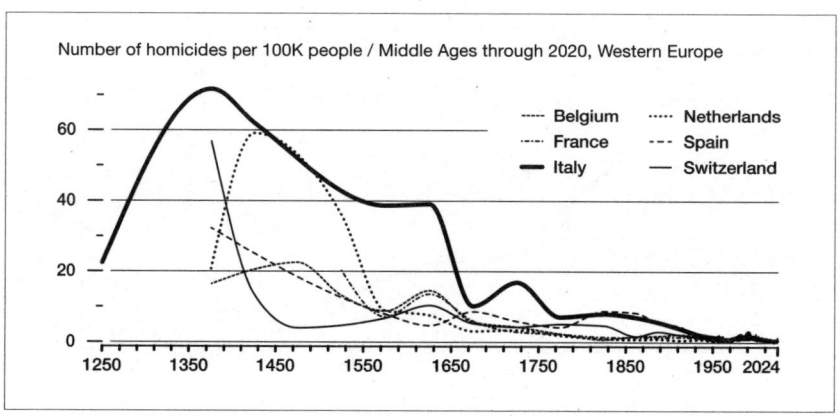

Homicide rates in Western Europe have collapsed over the past seven centuries. Medieval Europe saw murder rates dozens of times higher than today. As states consolidated power, legal systems matured, and cities built sanitation and policing networks, violence steadily lost its grip. Forensics, emergency medicine, and social trust reinforced the trend. The result is one of civilization's most underappreciated achievements: a world where routine murder has become rare, and peace, for most, is the default condition.

Chart 31
Reported Crime Rate in the U.S.

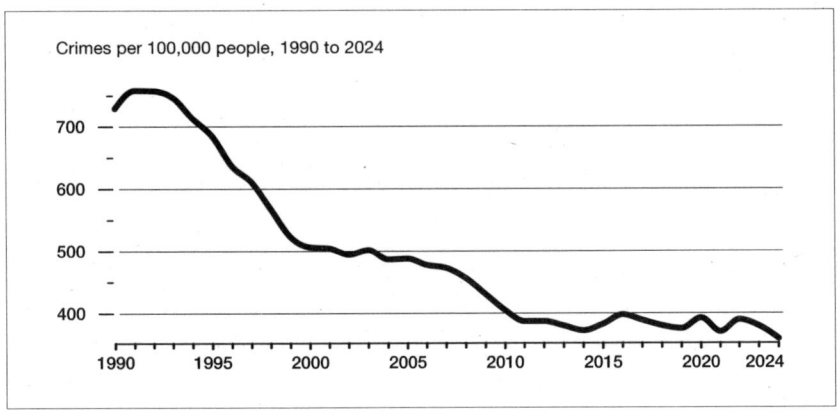

Since the early 1990s, crime in the United States has fallen dramatically, an enduring and often misunderstood trend. Both violent and property crimes have declined across nearly every major city, even as populations grew, technology disrupted work and life, and inequality widened. The reasons are layered: Better policing analytics, improved urban design, expanded economic opportunities, and tighter community networks all played roles. Environmental changes—from cleaner air to lead removal—also mattered. The pattern suggests that public safety can emerge not from heavier punishment or surveillance, but from smarter institutions and an ecosystem of civic and technological progress working in tandem.

APPENDIX B

Charts Showing the Dark Side of Abundance

Exponential technology has lifted billions from poverty, extended lifespans, and redefined what's possible for the species. But progress casts a long shadow. This appendix examines the other half of the story—the steep cost of abundance at scale.

Drawn from long-term datasets such as Our World in Data, these charts trace the negative feedback loops of modern progress. As carbon emissions rise with industrial output, ecosystems absorb the cost. Diets built on sugar and convenience have replaced hunger with metabolic disease. Hyperconnectivity has blurred the line between information and overload. Prosperity has expanded opportunity, but also anxiety, loneliness, and chronic stress.

The same forces that make life easier often make it harder to live well. Complexity is rising faster than our capacity to adapt. Attention fractures, ecosystems strain, and the climate edges past thresholds that once seemed unimaginable.

These aren't signs that progress has failed; they show where it's out of balance. The ingenuity that built abundance can also restore equilibrium. But if you've come through these pages wondering where you can make a difference, these charts indicate the pressure points of the century—where innovation meets responsibility.

Chart 32
Rising Carbon Threatens Our Oceans

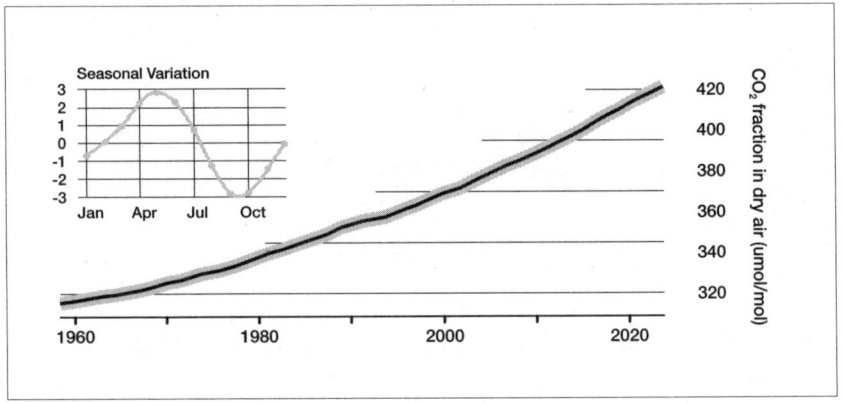

What began as the postwar industrial boom evolved into a vast, unplanned experiment in planetary chemistry. The same carbon that fueled global prosperity and technological progress has also destabilized the delicate balance of Earth's oceans. As atmospheric CO_2 climbed past 420 parts per million, the seas absorbed much of it. Rising acidity erodes coral reefs, weakens marine ecosystems, and alters food chains that sustain billions. This curve captures more than emissions; it maps the tension between human advancement and environmental resilience. The next era of abundance will depend on breaking this linkage—building regenerative energy systems that sustain both growth and the biosphere that makes it possible.

Chart 33
Obesity in the U.S. and the World

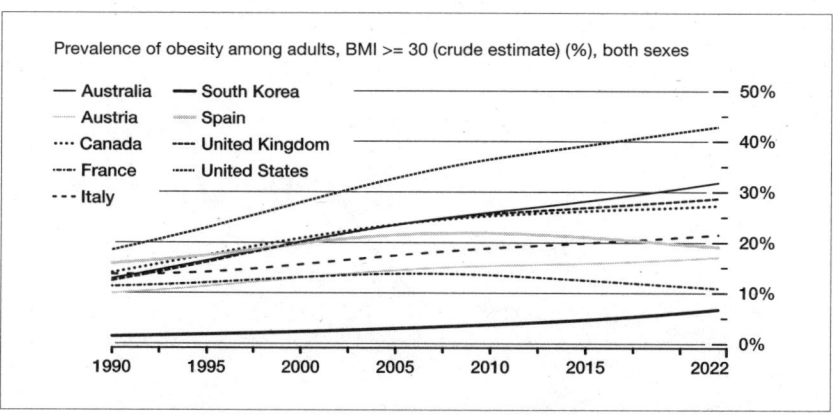

As calories became cheap, plentiful, and sweet, obesity surged across both developed and developing nations. The United States now counts over 40 percent of adults as obese, with similar trends emerging globally. Behind the numbers lie engineered foods, sedentary jobs, and the quiet domination of algorithmic convenience. What began as the victory of food security has turned into a complex systems failure. Reversing it will mean reimagining cities, diets, and incentives to reward vitality over consumption, redefining progress not as comfort, but as sustained health.

Chart 34
Sugar Consumption US/UK, 1700-2000

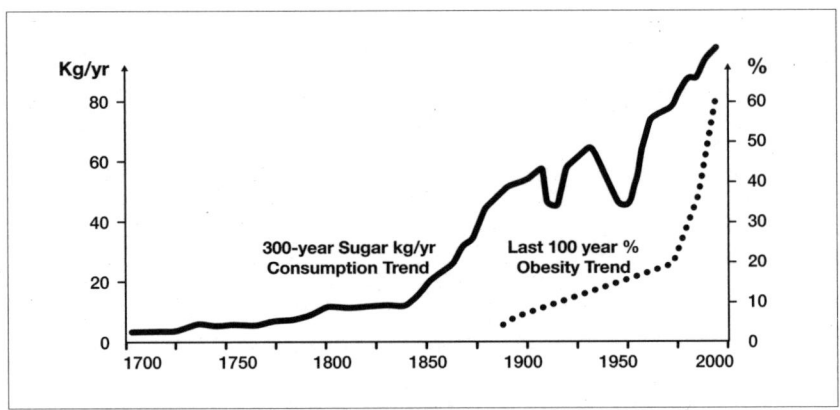

Sugar's story is a microcosm of modern progress: from rare indulgence to industrial staple. In 1700, per-capita sugar consumption in Britain and the United States was negligible; by 1900 it had multiplied more than twentyfold, fueled by colonial trade, mechanized refining, and global shipping. The twentieth century transformed sweetness into a constant background flavor of daily life, cheap, addictive, and omnipresent. When metabolic limits collided with modern abundance, obesity and diabetes surged. Reversing the trend will demand more than willpower—it will require redesigning economies, incentives, and culture around metabolic integrity, not excess.

Chart 35
Young People Are More Depressed

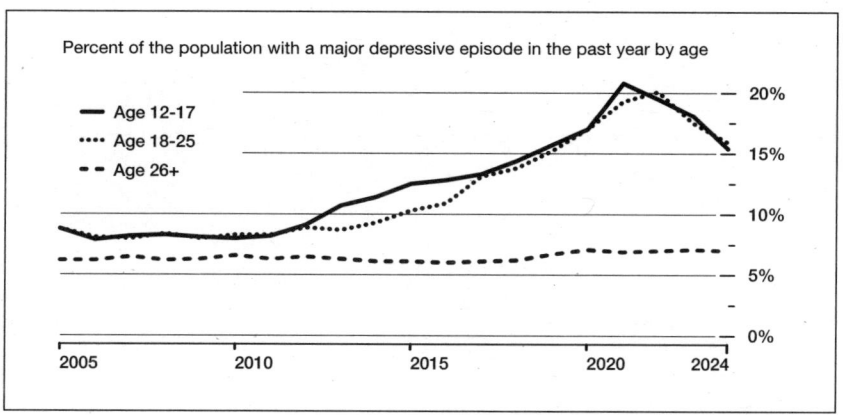

Rates of depression among young people have climbed sharply since the early 2010s, tracing a curve that mirrors the rise of the smartphone era. Constant connectivity has expanded social reach but hollowed real connection, replacing community with curation and attention with comparison. Adolescence—once a time of local belonging and physical exploration—has shifted online, where validation is quantified and anxiety amplified. Economic pressures, climate dread, and information overload compound the strain. The data point to a deeper challenge: building psychological resilience in an age where the tools meant to connect us often leave us feeling more alone.

Chart 36
Increase in Antibiotic Resistance

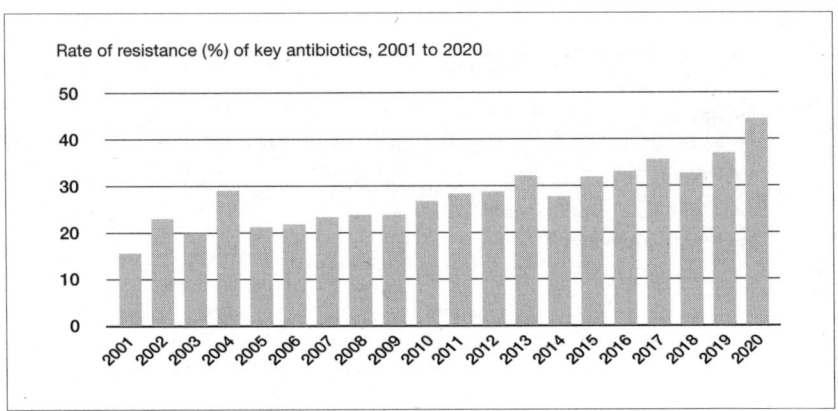

Resistance to once-reliable antibiotics has climbed steadily since the mid-twentieth century, driven by overuse in medicine and agriculture. Each dose that kills weak bacteria leaves behind the strong, accelerating evolution in hospitals and feedlots alike. The result: infections that no longer respond to standard treatments. Rapid DNA-based diagnostics, phage therapies, and synthetic biology are now racing to restore our advantage, but the bacteria are still evolving faster than the system regulating them.

Chart 37
Rapid Increase in Plastic Pollution

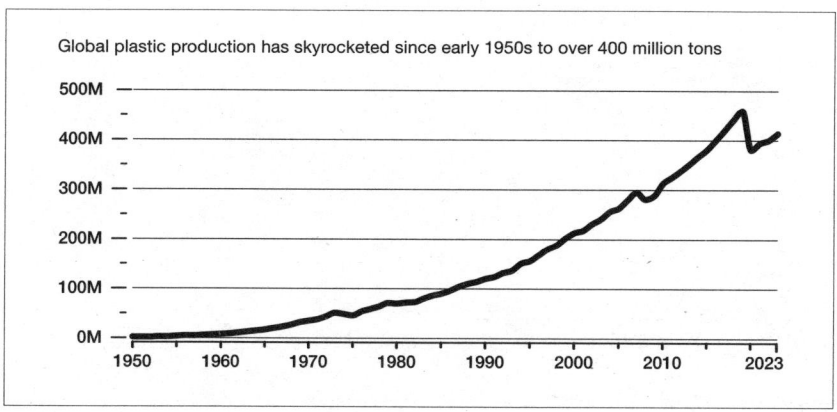

Global plastic production has surged from near zero in the 1950s to more than 400 million tons a year. What began as a cheap, durable marvel quickly became the foundation of modern convenience—and waste. The same traits that made plastics indispensable also made them persistent, filling oceans and bloodstreams alike. Microplastics are now found in human lungs, placentas, and even the brain, their long-term biological effects still poorly understood. In the environment, they alter soil structure, disrupt marine food webs, and leach hormone-mimicking chemicals into ecosystems. Innovation is now chasing its own tail: Enzymatic recyclers, bioplastics, and circular-economy policies are trying to clean up a century's worth of chemistry.

Chart 38
Electronic Waste (E-Waste)

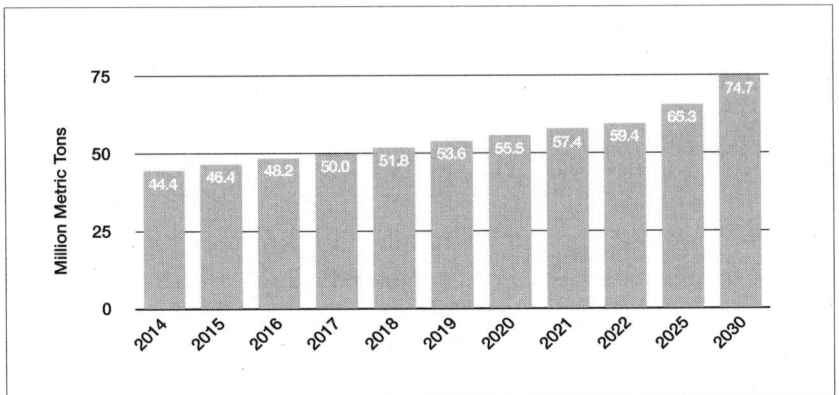

E-waste is the debris field of digital abundance. As devices get faster, sleeker, and cheaper, they've also become more disposable. Global electronic waste has grown from about 44 million metric tons in 2014 to nearly 60 million in 2022, on track to hit 75 million by 2030. The fallout is both environmental and human: Toxic metals like lead and cadmium contaminate soil and groundwater, while informal recycling exposes workers to dangerous fumes. Yet inside this waste lies a vast urban mine filled with gold, copper, and rare earths waiting to be recovered. Smarter product design, global repair standards, and closed-loop supply chains could make e-waste less of a crisis and more of a resource.

Chart 39
Greenhouse Gas Emissions

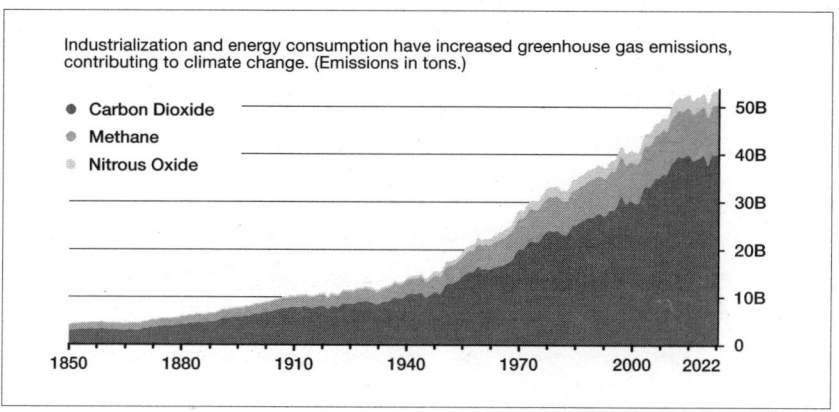

Carbon dioxide, methane, and nitrous oxide emissions have climbed in lockstep with industrial growth, tracing the atmospheric footprint of modern prosperity. Fossil fuels powered the revolutions in transport, agriculture, and manufacturing, but at the cost of destabilizing the climate system that underpins them. The challenge now is to break this coupling: electrify transport, decarbonize grids, and reengineer materials so that economic growth no longer tracks carbon output. The tools exist—solar, wind, nuclear, and AI-optimized energy systems—but the transition's pace will decide whether this century's innovation rewrites the climate curve or merely bends it.

Chart 40
Average Attention Span Is Shortening

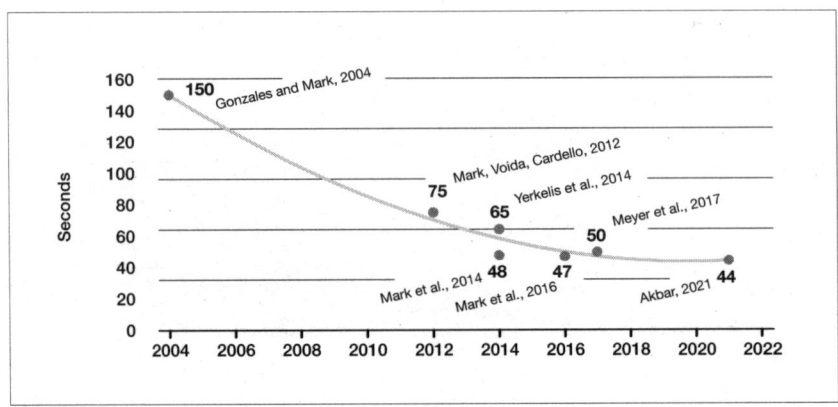

In 2004, people could focus on a single screen task for about 150 seconds. By 2021, that window had shrunk to just 44. The constant flicker of notifications and context switching has rewired attention itself, turning multitasking into microtasking. The effects reach beyond productivity: Fragmented focus correlates with higher stress and reduced creative depth. Still, attention is not extinct, only redistributed. The same systems that hijack it—algorithmic feeds, adaptive interfaces—could be redesigned to restore flow, training the brain to sustain depth rather than chase novelty.

APPENDIX C

Exponential Technology Driving Increased Abundance

Welcome to the engine room of the age of abundance: Charts that showcase the technologies transforming every sector, from AI and robotics to renewable energy, synthetic biology, and quantum computing. Appendix C is where possibility meets proof, illustrating how technology is not evolving linearly but accelerating exponentially, driving down costs, boosting performance, and making once-scarce resources universally plentiful.

Whether it's compute doubling every few months, genome sequencing becoming cheaper than a cup of coffee, or sensors and networks enabling billions of connected devices, abundance is being created at a record pace. These are the forces reshaping industries, rewiring economies, and redefining what it means to be human.

The charts also underpin the progress in appendix A and the challenges in appendix B. They are the technological bedrock of tomorrow, and a map to where the world is headed.

Whether you're a student, investor, entrepreneur, or policymaker, these insights are here to inform and inspire. One thing is certain: If we're going to solve the challenges ahead, we need all of us on the same page—and sooner rather than later.

For an expanded collection of these and other charts, visit www.Diamandis.com/AbundanceCharts.

Chart 41
U.S. Households with TV

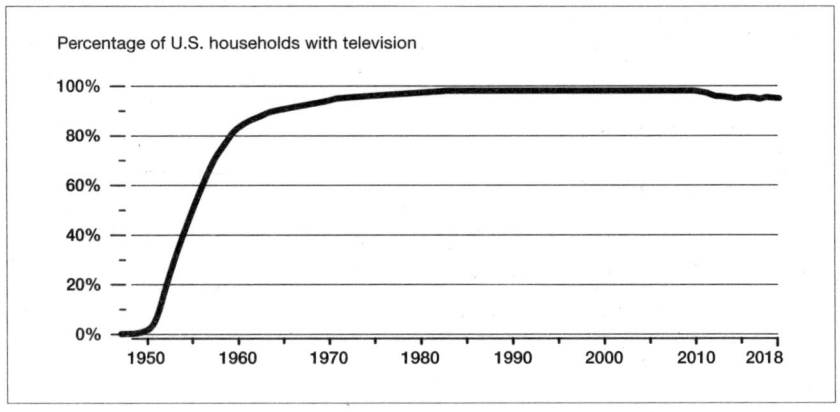

Television ownership in the United States went from nearly zero to almost universal in barely a decade—rising from 9 percent of households in 1950 to over 90 percent by 1960. No other medium spread faster. The TV's rapid ascent turned entertainment into infrastructure, uniting millions around synchronized nightly broadcasts. It marked the moment mass communication became a shared civic experience rather than a privilege of wealth.

Chart 42
Households with Computers

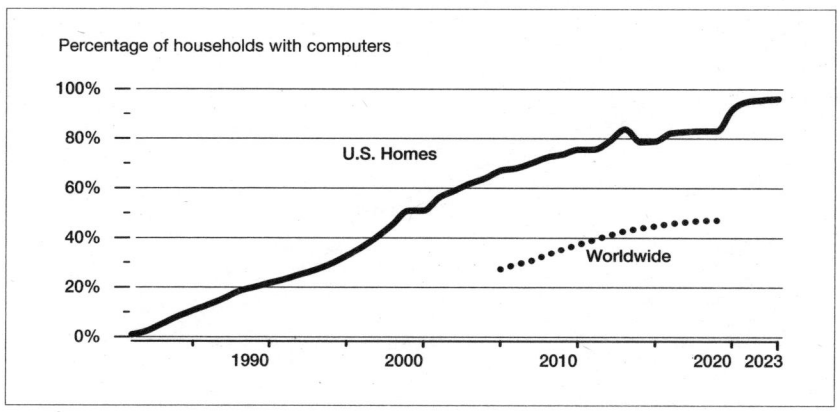

The rise of the personal computer transformed homes into information hubs. In the 1980s, only a small share of U.S. families owned one; by the 2000s, computers were everywhere, reshaping how people worked, learned, and connected. Falling hardware costs, better software, and easier interfaces turned the computer from luxury to necessity. The pattern repeated worldwide as access spread, proving that digital tools had become essential gateways to education, opportunity, and participation in the modern economy.

Chart 43
Falling Computing Costs

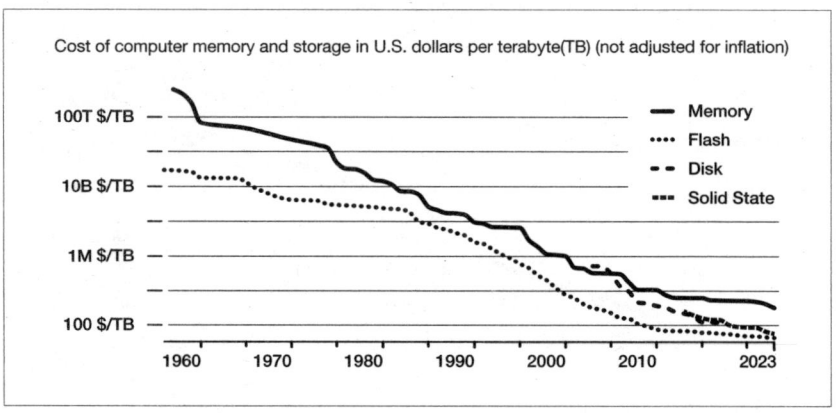

The cost of storing and processing data has collapsed by more than twelve orders of magnitude since the dawn of computing, from millions of dollars per terabyte in the 1960s to just a few cents today. This relentless deflation, driven by miniaturization, manufacturing advances, and global competition, turned computation from an institutional privilege into a universal utility. What once powered mainframes now fits in a pocket, and what once required budgets the size of nations now runs on cloud credit cards. As storage and memory approach near-zero cost, scarcity in the digital realm gives way to an era where intelligence, not hardware, becomes the true limit.

Chart 44
1 Sextillion X Improvement in Computation/$

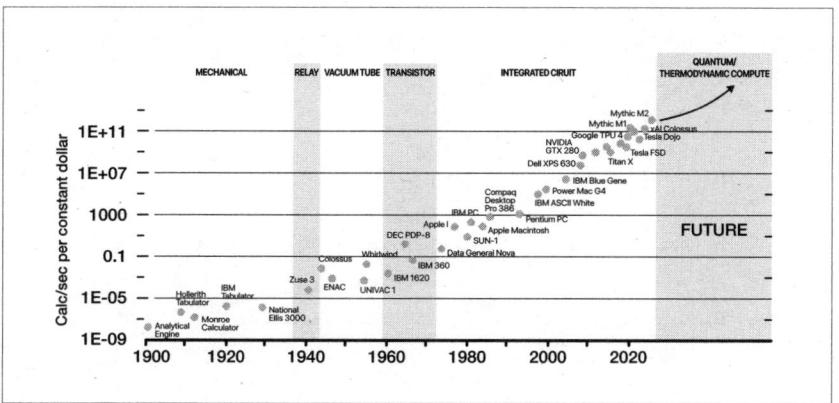

From the gears of Babbage's Analytical Engine to today's neural accelerators, computing power per dollar has risen by roughly one sextillion times. Each leap—mechanical, relay, vacuum tube, transistor, integrated circuit—reset the curve rather than flattening it. Even as silicon nears its physical limits, new frontiers like quantum and thermodynamic computing are extending the exponential, promising gains once thought unreachable.

Chart 45
The Cost Per Million Tokens Has Fallen Quickly

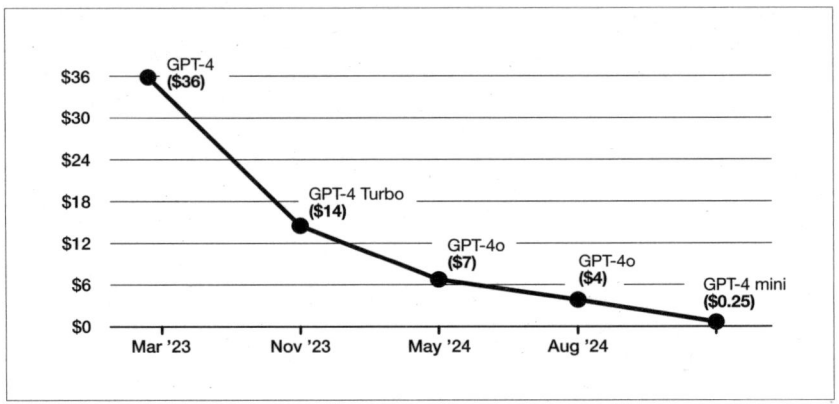

Artificial intelligence costs plummeted faster than any technology in history, making sophisticated AI available to small developers and individuals within months of its initial release. In 2023, accessing GPT-4 cost roughly $36 per million tokens. By mid-2024, GPT-4o mini brought that below twenty-five cents. The price of intelligence is collapsing even faster than the price of computation once did, signaling a new kind of abundance: cognitive rather than mechanical.

Chart 46
Global Data Creation Is About to Explode

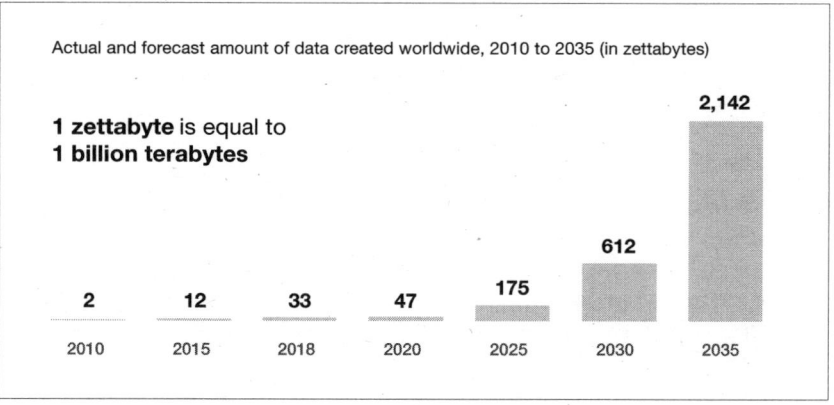

Global data creation is compounding at a staggering rate, from 2 zettabytes in 2010 to a projected 2,000-plus by 2035. Every connected sensor, camera, satellite, and transaction adds to an expanding record of the planet's activity. The result is a near-continuous digital reflection of the physical world, where almost everything is tracked, stored, and analyzed. This explosion marks the end of informational scarcity: Insight now depends less on data access than on interpretation. As machine intelligence mines these oceans of information, prediction begins to rival observation, and the world becomes measurable in real time.

Chart 47
Compute Increasing 4.6x per Year

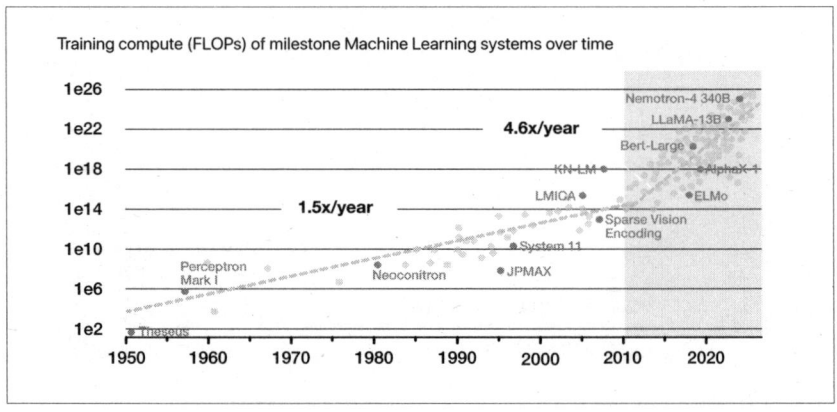

AI's appetite for compute has grown faster than almost any other technological demand in history, now doubling roughly every six months, or about 4.6x per year. What began in the 1950s with systems like the Perceptron has exploded into trillion-parameter models that strain even the largest data centers. Each leap in capability—from vision recognition to multimodal reasoning—has been powered by more computation, driving innovations in GPUs, tensor processors, and distributed training at planetary scale. The feedback loop is self-accelerating: Greater compute enables smarter models, which in turn design better chips and algorithms. Intelligence is no longer capped by theory but by energy, hardware, and imagination.

APPENDIX C

Chart 48
Hyperscaler Capex By Year

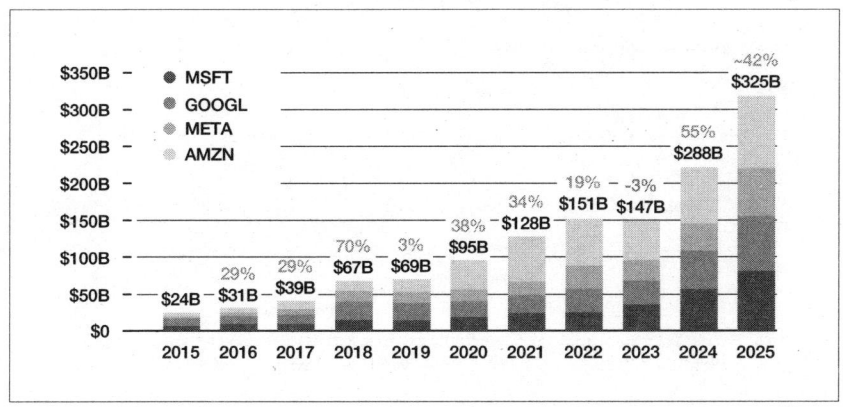

Capital spending by hyperscalers—Microsoft, Google, Meta, and Amazon—has surged from $24 billion in 2015 to more than $325 billion projected in 2025. No private investment wave in history has built more computational capacity. These data centers now operate as the backbone of the modern digital world, supporting everything from AI training to real-time logistics and global communication.

Chart 49
Automobile & Airline Fatality Rates

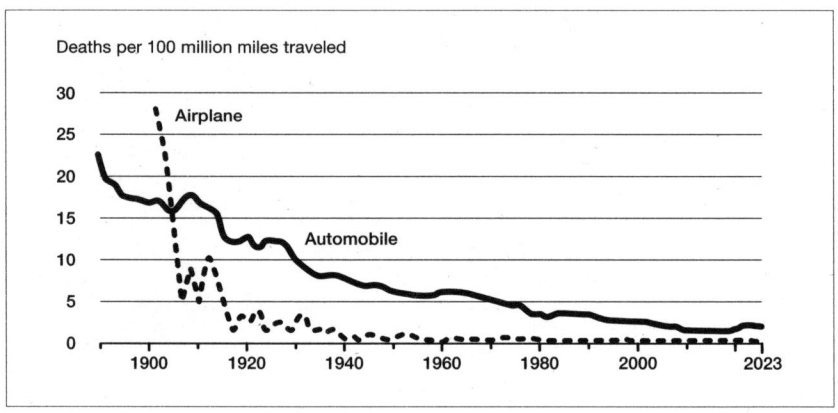

Over the past century, travel has become exponentially safer. Early motorists and aviators faced double-digit death rates per hundred million miles; today, both are near zero. The decline reflects the compounding effects of engineering rigor, regulatory reform, and automation, from seat belts to radar to AI-assisted control systems. Aviation's steep safety gains show what happens when technology and oversight evolve together, while automotive fatalities continue to fall as autonomy advances. Risk hasn't disappeared, but it's been engineered down to the margins.

Chart 50
Robotaxi Vs. Uber, Cost per Mile

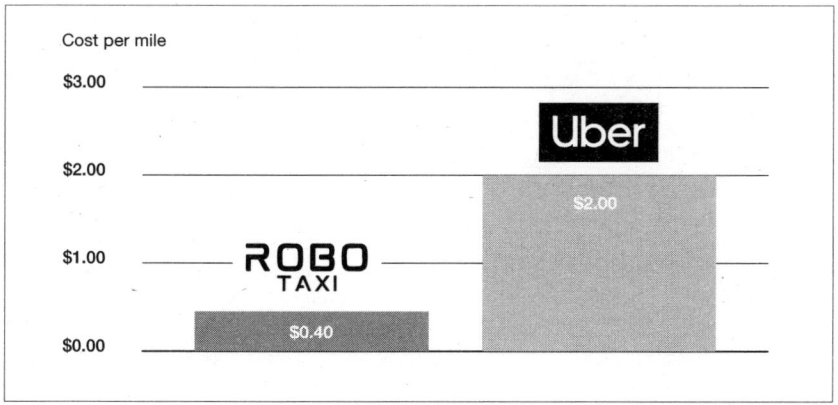

Robotaxis are projected to operate at roughly $0.40 per mile—about one-fifth the cost of a human-driven Uber. The savings come almost entirely from removing labor, the largest expense in transportation. By automating driving, AI drives down costs while maintaining service quality, making on-demand travel far more affordable and accessible. Unlike past transport revolutions that took decades to lower prices, autonomy delivers cost reductions from the start. As these services expand, they could redefine how people move through cities and rethink what it means to own a car.

Chart 51
Cost of Space Travel Plummets

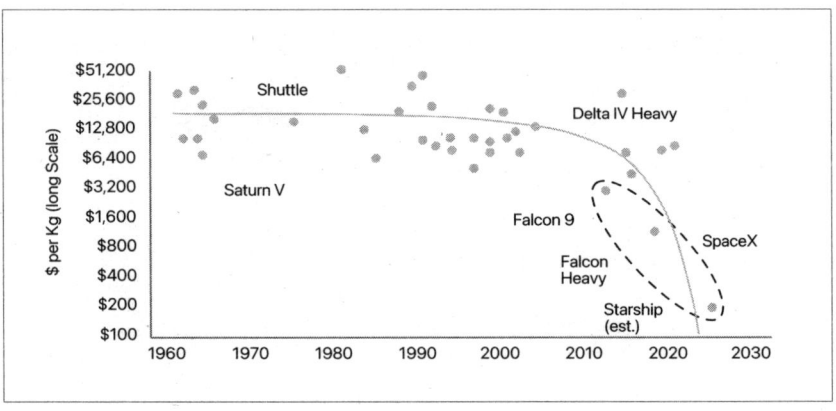

Space launch costs have fallen from tens of thousands to just hundreds of dollars per kilogram, thanks to reusable rockets and new manufacturing methods. After half a century of stagnation, this sudden drop shows that breakthrough engineering can overcome seemingly permanent technical barriers. Private competition achieved in a decade what government programs couldn't deliver through fifty years of gradual progress. The result: Space is shifting from an exclusive domain to an open frontier, powering everything from global connectivity to commercial missions once deemed impossible.

Chart 52
Exponential Increase in Objects in Space

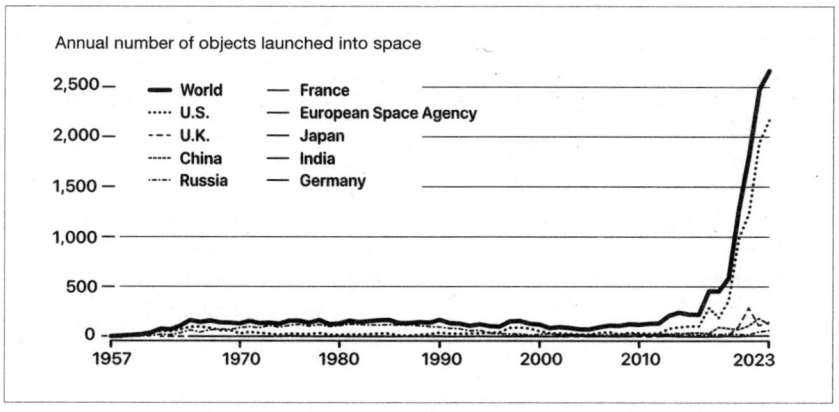

For half a century, the cost of reaching orbit barely budged, then SpaceX rewrote the economics. Launch prices that once hovered above $20,000 per kilogram have plunged below $500, driven by reusable boosters, mass manufacturing, and relentless iteration. The drop transformed space from a government monopoly into a functioning market. Private firms now deploy satellites, ferry cargo, and plan deep-space missions at a fraction of Cold War costs. What NASA once achieved with national budgets, start-ups now approach with venture rounds.

Starship represents the next inflection: an orbital freighter designed to drive costs below $100 per kilogram and carry over 100 metric tons per launch. If successful, it will enable on-orbit construction, lunar industry, and the first true supply chain between Earth and Mars.

Chart 53
Sequencing the Human Genome

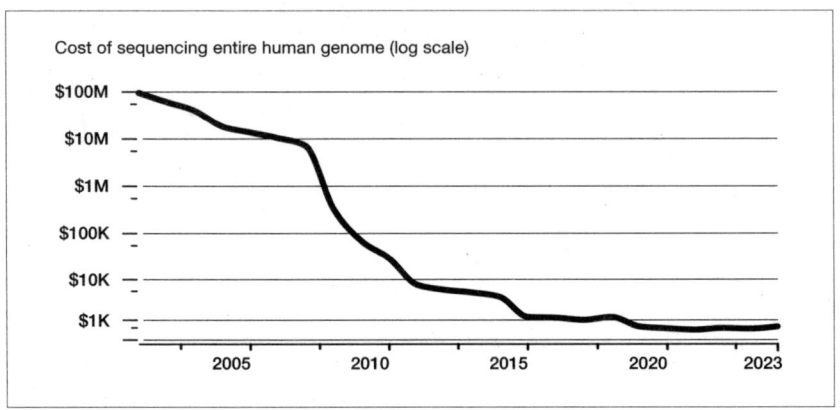

Sequencing a human genome once cost billions; today it's measured in hundreds of dollars. The curve fell faster than Moore's law, powered by automation, high-throughput chemistry, and machine learning. What began as a decade-long international effort is now a routine lab procedure. This collapse in cost has moved genomics from elite science to everyday medicine, enabling personalized treatments, rapid pathogen tracking, and population-scale health studies.

Chart 54
Massive Growth in China Solar PV

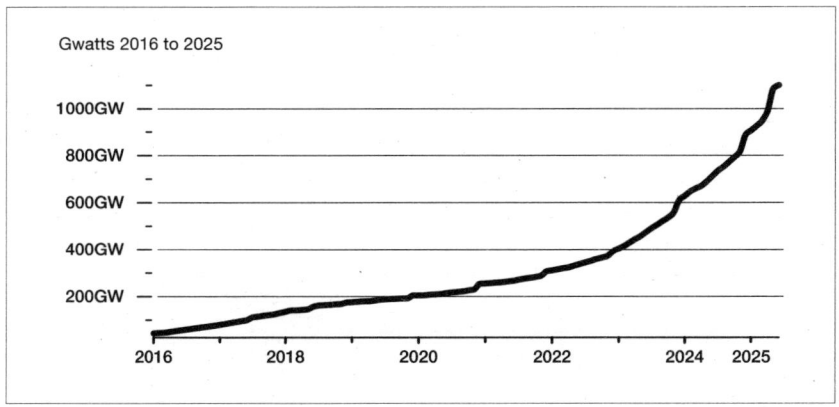

China's solar capacity has exploded from under 200 gigawatts in 2016 to nearly a terawatt in 2025, a record-breaking expansion of clean energy capacity. This surge is more than national energy policy; it's a global supply chain event. By scaling production of photovoltaic cells and components, China has driven the cost of solar power to record lows worldwide, turning sunlight into the cheapest source of new electricity on Earth. The curve shows the power of manufacturing at scale: Every panel installed makes the next one cheaper. Energy abundance is no longer theoretical. It's being mass-produced.

Chart 55
Worldwide Renewable Electricity Generation

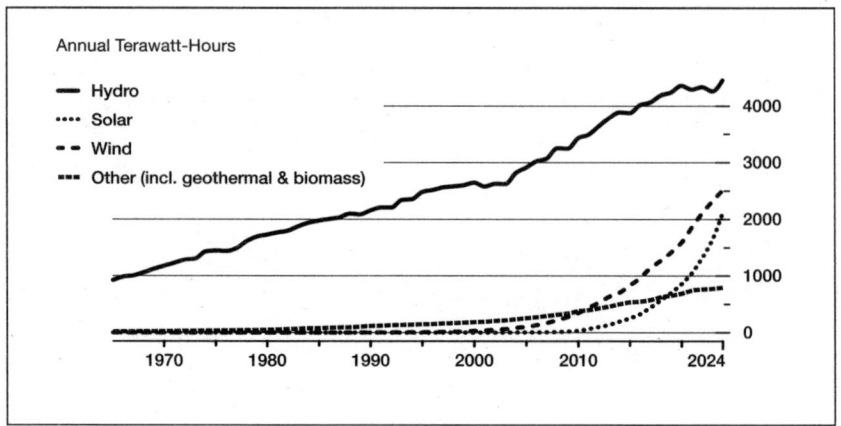

Renewable electricity has shifted from niche experiment to global mainstay in less than two decades. Solar and wind now add capacity faster than any energy source in history, their growth curves resembling those of computing more than coal. The transition overturns a century of assumptions about energy inertia—technologies once dismissed as supplemental now anchor national grids. Unlike fossil fuels, renewables scale with innovation rather than deplete with use. The result is an energy system rewired around abundance, where progress is limited not by resources but by imagination and storage.

Chart 56
Battery Cost Continues to Drop

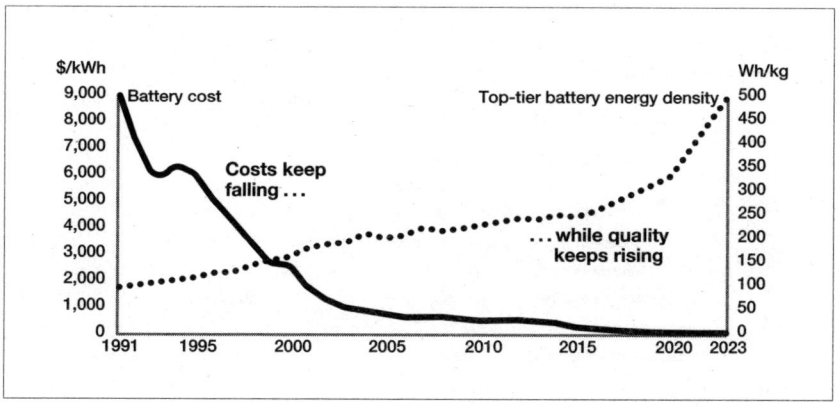

Battery prices have fallen from over $9,000 per kilowatt-hour in the early 1990s to well under $150 today, even as energy density has more than doubled. The result is a rare dual curve, with costs plummeting while performance soars. This dynamic mirrors Moore's law, but for physical energy systems, turning batteries from niche components into core infrastructure for transport and the grid. Cheap, high-density storage eliminates the biggest constraint on renewables: intermittency. When energy can be stored as easily as data, electricity stops being weather dependent and becomes programmable, making energy efficiency available on demand.

Chart 57
Private Sector Fusion

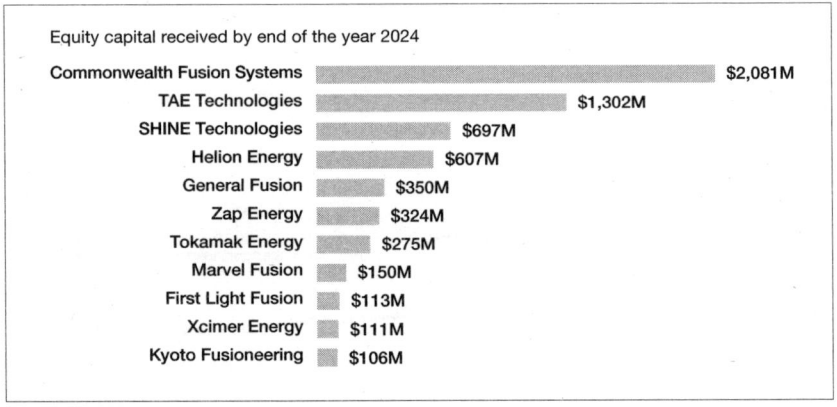

Private investment in fusion has exploded, led by start-ups like Commonwealth Fusion Systems, TAE Technologies, and Helion Energy. After decades dominated by government labs chasing scientific proof, the focus has shifted to commercialization and engineering speed. Venture-backed fusion start-ups now promise grid-scale reactors within the 2030s, treating the physics as solved and the challenge as one of design. This is the Silicon Valley model applied to energy's ultimate frontier: clean, limitless power from the most abundant fuel in the universe. If even one succeeds, the era of energy scarcity ends.

ACKNOWLEDGMENTS

Peter's Acknowledgments

I would like to acknowledge and deeply thank my mentor, Ray Kurzweil, for helping me to shape and adopt my mindset and hunger for all things exponential.

Equal gratitude goes to my writing partner, Steven Kotler, whose godlike writing and storytelling skills will surpass all future superintelligence.

I would like to thank a few key people on my PHD Ventures, Abundance, and research team for their support in the research and production of this book.

Cheo Rose-Washington and Max Song for their brilliant research support. Esther Count and Julie van Amerongen for their incredible leadership of my Abundance ecosystem. Joel Cave, whose design genius supported the creation of our appendix and the cover. Finally, to Tyler Donahue, Nick Singh, Danna Kahn, and Teddy Garcia for their marketing miracles.

Finally, gratitude to my personal board of directors: Louis G. Reese IV, Keith Ferrazzi, Eric Pulier, Salim Ismail, and Marcus Shingles.

Steven's Acknowledgments

I would like to acknowledge and deeply thank Mihaly Csikszentmihalyi for founding the field of flow science and being a friend. We miss you.

Equal gratitude goes to my writing partner, Peter Diamandis, for thirty years of friendship, laughter, brilliant ideas, fantastic words, and a huge heart that will not quit.

I also need to thank a few key people on my amazing team. Michael Wharton, without whom all would surely be lost. Michael Mannino for all things neuro and, mos def, for *Lose Yourself*. Ryan Wickes for all things dangerous and, especially, Wednesdays at the Wood. Kelsey Wyman for helping me juggle chain saws. Chase Adams for keeping the lights on. Rian Doris, because of the ride, and landing the plane.

Finally, gratitude to my personal board of wise miscreants: Jody Levy, Sarah Sarkis, Rosie Acosta, Vika Victoria, Keoki Flagg, Joshua Lauber, and, of course, Kiko.

Peter and Steven's Acknowledgments

Finally, Peter and Steven are forever grateful for Kristen Diamandis who once again (as she did with *Abundance*) designed a powerful and meaningful book cover. We also want to thank our agents, John and Max Brockman, and everybody at Simon & Schuster—especially our editor, Stephanie Frerich—for all the hard work and tireless support.

Lastly, to our communities—all of you at A360, FRC, SU, XPRIZE—thanks for believing.

NOTES

PART 1

Chapter 1: Theogony

3 *"We are as gods"*: Stewart Brand, *The Whole Earth Catalog*, Fall 1968. It first appeared on the title page of the *Whole Earth Catalog*'s inaugural issue, published in Menlo Park, California.

4 *Total Old Testament Miracles: 83:* The enumeration of eighty-three miracles in the Old Testament, organized into ten categories, is a synthesis drawn from multiple scholarly interpretations and compilations, including Wilfred Graves Jr.'s "List of Miracles in the Old Testament," Wilfred Graves Ministries, July 2019, https://www.wilfredgraves.org/wp-content/uploads/2019/07/List-of-Miracles-Old-Testament.pdf. While exact counts and classifications may vary depending on interpretive lens or denominational tradition, this structure is representative of the diversity and scope of divine interventions documented in the Hebrew Bible.

5 *"I was blind, but now I see"*: John 9:25 (ESV): "He answered, 'Whether he is a sinner I do not know. One thing I do know, that though I was blind, now I see.'"

6 *Hodak's primary focus:* Zoë Corbyn, "Max Hodak: 'Brain-Machine Interfaces Are About Creating a New Symbiosis with Machines,'" *The Guardian*, September 16, 2023, https://www.theguardian.com/science/2023/sep/16/max-hodak-interview-science-corp-neuralink.

6 *Technically, Neuralink is a triple miracle:* Neuralink's stated mission is to develop brain-computer interfaces that restore autonomy to people with neurological conditions by enabling direct control of devices through thought. See: "Neuralink," https://neuralink.com. The company has demonstrated

initial success in allowing a paralyzed individual to control a computer cursor with his mind. These interfaces potentially enable what could be described as "modern telepathy" (brain-to-brain communication) and "telekinesis" (thought-based control of machines). See: Andrew Griffin, "Elon Musk's Neuralink Shows Brain Chip Working in Human for First Time," *The Independent*, January 30, 2024, https://www.independent.co.uk/tech/neuralink-brain-chip-human-test-b2484482.html.

6 *PRIMA retinal prosthetic:* Science Corp White Paper, 2023, https://science.xyz/technologies/prima/.

6 *Twenty million people suffer:* Spinal cord injury numbers compiled by the National Spinal Cord Injury Statistics Center, see: "Traumatic Spinal Cord Injury Facts and Figures at a Glance," NSCISC, 2022, https://msktc.org/sites/default/files/SCI-Facts-Figs-2022-Eng-508.pdf.

6 *affects over 170 million people:* Macular degeneration stats, see: Eileen Bailey, "Age-Related Macular Degeneration Expected to Affect 288 Million People by 2040," *Medical News Today*, January 11, 2024, https://www.medicalnewstoday.com/articles/age-related-macular-degeneration-expected-to-affect-288-million-people-by-2040#:~:text=The%20researchers%20determined%20that%20age,may%20not%20actually%20be%20increasing.

6 *normal vision is 20/20:* "Low Vision and Legal Blindness Terms and Descriptions," American Foundation for the Blind, https://www.afb.org/blindness-and-low-vision/eye-conditions/low-vision-and-legal-blindness-terms-and-descriptions?gad_source=1&gad_campaignid=1979223497&gbraid=0AAAAADnSdap44z_SWyg6x3lLVHHmcl1Ez&gclid=CjwKCAjw87XBBhBIEiwAxP3_A7gdI3a05QDGWc5L5cHSJtl220ftzoY0AynaLonvwD46dH2-SN6EGRoCCUIQAvD_BwE.

7 *eyesight improved to 20/160:* José-Alain Sahel et al., "Partial Recovery of Visual Function in a Blind Patient After Optogenetic Therapy," *Nature Medicine* 27, no. 7 (2021): 1223–29, https://doi.org/10.1038/s41591-021-01351-4.

7 *Moore's law is the standard example:* Gordon E. Moore, "Cramming More Components onto Integrated Circuits," *Electronics* 38, no. 8 (1965); reprinted by Intel at https://www.intel.com/content/www/us/en/newsroom/resources/moores-law-electronics-magazine-1965.html.

8 *one-hundred-fold gains:* Moore's law originally described a twofold increase in transistor density every eighteen to twenty-four months. Between 2012 and 2022, computational power—particularly in AI-specific hardware—accelerated well beyond that rate. For example, AI training compute grew by a factor of ten times per year from 2012 to 2018, as documented in: OpenAI, "AI and Compute," May 16, 2018, https://openai.com/research/ai-and-compute. By 2023–2024, the rise of GPU-driven generative AI (e.g.,

Nvidia H100 and custom TPU architectures) pushed computational scaling toward one hundred times per year in some domains. See: Jensen Huang, "The Era of Generative AI," GTC Keynote, Nvidia, "The Era of Generative AI," GTC Keynote by Jensen Huang, March 2024; Chris Re et al., "Scaling Laws for Foundation Models," *Stanford CRFM Report*, 2023, https://crfm.stanford.edu/2023/ScalingLaws.

9 *over thirty autonomous car companies:* "Autonomous Driving's Future: Convenient and Connected," McKinsey, January 6, 2023, https://www.mckinsey.com/industries/automotive-and-assembly/our-insights/autonomous-drivings-future-convenient-and-connected.

9 *robots running their warehouses:* "Warehouse Automation Trends," CB Insights, 2023.

9 *in the Middle East and Asia:* "Flying Taxis Are Coming to the Middle East," Bloomberg, January 2024.

10 *cognitive scientist Dedre Gentner:* Dedre Gentner, "Structure-Mapping: A Theoretical Framework for Analogy," *Cognitive Science* 7 (1983): 155–70, https://doi.org/10.1016/S0364-0213(83)80009-3.

11 *intelligence and creativity:* Douglas Hofstadter and Emmanuel Sander, *Surfaces and Essences: Analogy as the Fuel and Fire of Thinking* (Basic Books, 2013).

11 *In Jung's definition:* Carl G. Jung, *Collected Works of C. G. Jung*, trans. R. F. C. Hull, vol. 9, part 1, *The Archetypes and the Collective Unconscious* (Princeton University Press, 1981).

12 *sixty films and thirty television series:* There is, of course, both a superhero database (superherodb.com) and a Wiki that tracks superhero films per decade: Wikipedia, "Category: Superhero Films by Decade," last modified February 15, 2019, https://en.wikipedia.org/wiki/Category%3ASuperhero_films_by_decade?utm_source=chatgpt.com.

12 *233 percent faster:* How much faster is life today than it was in 2010? We built a model to find out. First, we picked ten big tech shifts that have redefined the infrastructure of daily life—things like internet access, smartphone adoption, AI, cloud computing, and how much data the world generates each year. Then we added ten major cultural changes: the legalization of same-sex marriage, the rise of streaming and remote work, greater mental health awareness, trans rights protections, and the explosion in daily digital content exposure.

Each variable was normalized on a 0–1 scale, with 2023 set to 1.0. Then we gave each one a weight from 1 to 10 based on how many people it impacts and how deeply it's changed the way we live. Internet use and smartphone ownership got the highest weights—10 and 9—since they affect billions of people. Genome sequencing, which is hugely important but still relatively niche, got a 2. Same idea on the cultural side: The spread of mental health awareness and remote work matter a lot, but the fact that over five billion people now live online got top billing.

Back in 2010, the weighted tech score came out to 0.259, which jumped to 1.0 in 2023. That's a 286 percent increase. The weighted cultural score started at 0.359 and also hit 1.0 by 2023—a 179 percent increase. If you average these together, we find that life in 2023 is moving 233 percent faster than it was in 2010.

Scope-Weighted Variables Driving Pace of Life Acceleration (2010–2023)

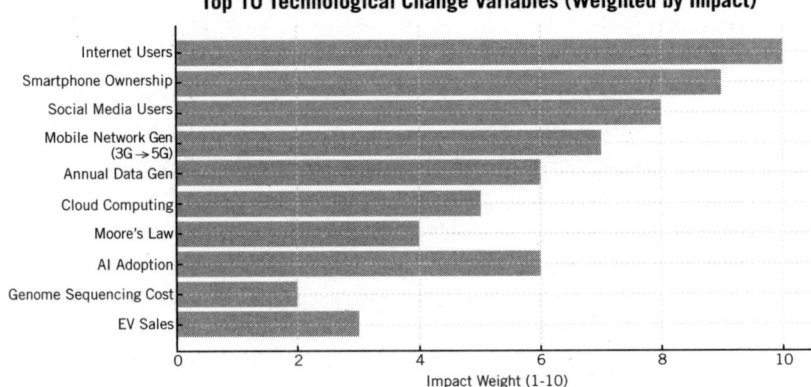

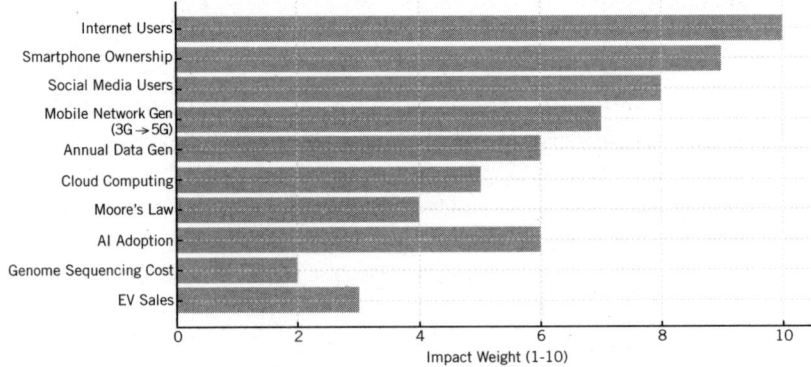

© Kotler/Diamandis 2026

17 *seen significant gains:* "The 2025 AI Index Report," Stanford University Human-Centered Artificial Intelligence, https://hai.stanford.edu/ai-index/2025-ai-index-report.

17 *a trillion dollars a year:* "Mental Health at Work," World Health Organization, September 2, 2024, https://www.who.int/news-room/fact-sheets/detail/mental-health-at-work.

18 *The brain is a prediction engine:* The idea that the brain is a prediction engine originates in predictive coding theory and the free energy principle, which propose that perception, cognition, and action serve to minimize prediction error and conserve metabolic energy. See: Karl Friston, "The Free-Energy Principle: A Unified Brain Theory?," *Nature Reviews Neuroscience* 11, no. 2 (2010): 127–38, https://doi.org/10.1038/nrn2787; Andy Clark, "Whatever Next? Predictive Brains, Situated Agents, and the Future of Cognitive Science," *Behavioral and Brain Sciences* 36, no. 3 (2013): 181–204, https://doi.org/10.1017/S0140525X12000477; Jakob Hohwy, *The Predictive Mind* (Oxford University Press, 2013); Lisa Feldman Barrett, *How Emotions Are Made: The Secret Life of the Brain* (Houghton Mifflin Harcourt, 2017); and Anil K. Seth, "*The Cybernetic Bayesian Brain: From Interoceptive Inference to Sensorimotor Contingencies,*" University of Sussex (2015), https://www.pensierocritico.eu/files/TheCyberneticBayesianBrain-PerceptualPresenceintheKuhnianPopperianBayesianBrain-InferencetotheBestPrediction-.pdf.

20 *four thousand books:* For estimates of historical information volume, see Martin Hilbert and Priscila López, "The World's Technological Capacity to Store, Communicate, and Compute Information," *Science* 332, no. 6025 (2011): 60–65, https://doi.org/10.1126/science.1200970; see also Clay Shirky, *Cognitive Surplus: Creativity and Generosity in a Connected Age* (Penguin Press, 2010).

20 *biggest information surge in history:* For data estimates and historical comparisons, see Hilbert and López, "Volume of Data Created, Captured, Copied, and Consumed Worldwide from 2010 to 2025"; "Volume of Data/Information Created Captured, Copied, and Consumed Worldwide From 2010 to 2019," *Statista*, November 19, 2025, https://www.statista.com/statistics/871513/worldwide-data-created/.

21 *Attention is maxed:* For foundational estimates, see Daniel Kahneman, *Attention and Effort* (Prentice-Hall, 1073); and Mihaly Csikszentmihalyi, *Flow: The Psychology of Optimal Experience* (Harper Perennial Modern Classics, 2008).

21 *two and a half hours a day:* Simon Kemp, "The Time We Spend on Social Media," DataReportal, January 31, 2024, https://datareportal.com/reports/digital-2024-deep-dive-the-time-we-spend-on-social-media.

21 *When a bee forages for nectar:* David A. Lawson, Heather M. Whitney, and Sean A. Rands, "Nectar Discovery Speeds and Multimodal Displays: Assessing Nectar Search Times in Bees with Radiating and Non-Radiating Guides,"

Evolutionary Ecology 31, no. 6 (2017): 899–912, https://doi.org/10.1007/s10682-017-9916-1.

21 *Yet heuristics don't always work:* For discussion of heuristics and biases, see Daniel Kahneman, Paul Slovic, and Amos Tversky, eds., *Judgment Under Uncertainty: Heuristics and Biases* (Cambridge University Press, 1982); and Daniel Kahneman, *Thinking, Fast and Slow* (Farrar, Straus and Giroux, 2011).

Chapter 2: An Abundance of Abundance

24 *Archeological studies suggest:* Dilman Mohammed Sabir, "Irrigation System in Ancient Mesopotamia," *Athens Journal of History* 11, no. 1 (2025): 73–88, https://www.athensjournals.gr/history/2025-11-1-4-Sabir.pdf.

24 *around 3500 BCE:* Charles Recknagel, "Were Horses First Domesticated in Kazakhstan?," Radio Free Europe / Radio Liberty, March 7, 2009, https://www.rferl.org/a/Horses_First_Domesticated_Kazakhstan/1505978.html.

24 *Genetic evidence shows:* Pablo Librado et al., "The Origins and Spread of Domestic Horses from the Western Eurasian Steppes," *Nature* 598 (2021): 634–40, https://doi.org/10.1038/s41586-021-04018-9.

24 *Trade, exploration, migration, culture:* "Transportation in Ancient Mesopotamia: Horses, Kunga, Carts and Boats," Facts and Details, accessed September 7, 2025, https://africame.factsanddetails.com/article/entry-1024.html.

24 *Vietnam's Mekong Delta:* Sonal Gupta, Victoria Mann, J. J. Mazzucotelli, Hafsa Maqsood, and Michelle Gomez, "Vietnamese Farmers Use Smartphones to Monitor Water Levels in Rice Fields," Mongabay, World Economic Forum, July 26, 2023; https://www.weforum.org/stories/2023/07/vietnam-water-scarcity-rice-farming-technology/.

25 *Crop yields stayed the same:* Gupta, Mann, Mazzucotelli, Maqsood, and Gomez, "Vietnamese Farmers Use Smartphones to Monitor Water Levels in Rice Fields."

26 *over two hundred million people:* Joe Hasell, Bertha Rohenkohl, Pablo Arriagada, Esteban Ortiz-Ospina, and Max Roser, "Poverty," Our World in Data, 2022, https://ourworldindata.org/poverty.

26 *More than a billion people:* "Energy Overview," World Bank, 2025, https://www.worldbank.org/en/topic/energy/overview.

26 *gained access to safe drinking water:* UN-Water, "Summary Progress Update 2021: SDG 6—Water and Sanitation for All," United Nations, July 2021, https://www.unwater.org/sites/default/files/app/uploads/2021/12/SDG-6-Summary-Progress-Update-2021_Version-July-2021a.pdf.

26 *In 2012, 2.4 billion people were online:* "Percentage of Global Population Accessing the Internet from 2005 to 2024, by Market Maturity," Statista, chart, May 31, 2024, https://www.statista.com/statistics/209096/share-of-internet-users-worldwide-by-market-maturity/.

26 *By 2025, the number exceeded 5.5 billion:* International Telecommunication Union, "ICT Statistics—Use of the Internet," ITU, accessed 2025, https://www.itu.int/en/ITU-D/Statistics/pages/stat/default.aspx.

26 *In 2012, a billion people owned a smartphone:* "Worldwide Smartphone Users Top 1 Bn, Says New Report," Phys.org, October 17, 2012, https://phys.org/news/2012-10-worldwide-smartphone-users-bn.html.

26 *By 2025, the number was over seven billion:* Shubham Singh, "iPhone Users & Sales Statistics 2025 (Worldwide Data)," DemandSage, September 15, 2025, https://www.demandsage.com/iphone-user-statistics/.

26 *remaking manufacturing and transportation:* Karen Hao, "A New Generation of AI-Powered Robots Is Taking over Warehouses," *MIT Technology Review*, August 6, 2021, https://www.technologyreview.com/2021/08/06/1030802/ai-robots-take-over-warehouses.

26 *As venture capitalist Vinod Khosla pointed out:* Vinod Khosla, "A Roadmap to AI Utopia," Khosla Ventures, November 11, 2024, https://www.khoslaventures.com/posts/a-roadmap-to-ai-utopia.

30 *So Musk got to work:* "Falcon 9," SpaceX, accessed September 7, 2025, https://www.spacex.com/vehicles/falcon-9.

30 *The results slashed costs tenfold:* "How Much Does It Cost to Launch a Reused Falcon 9? Elon Musk Explains Why Reusability Is Worth It," ElonX.net, September 19, 2020, https://www.elonx.net/how-much-does-it-cost-to-launch-a-reused-falcon-9-elon-musk-explains-why-reusability-is-worth-it/.

31 *So Musk checked prices:* Tesla, "Impact Report 2020," 7–28, https://www.tesla.com/ns_videos/2020-tesla-impact-report.pdf.

31 *Start with energy:* National Renewable Energy Laboratory, "Solar Energy Basics," updated August 27, 2025, https://www.nrel.gov/research/re-solar.

31 *Water scarcity follows a similar pattern:* "How Much Water Is in the Ocean? About 97 Percent of Earth's Water Is in the Ocean," National Ocean Service, last updated June 16, 2024, https://oceanservice.noaa.gov/facts/oceanwater.html.

31 *mega-ocean buried four hundred miles:* "New Evidence for Oceans of Water Deep in the Earth," Brookhaven National Laboratory, June 13, 2014, https://www.bnl.gov/newsroom/news.php?a=111648.

31 *97.5 percent of our surface water:* "Why Is the Ocean Salty?," United States Geological Survey, last reviewed March 14, 2022, https://www.usgs.gov/faqs/why-ocean-salty.

31 *solar-power desalination costs:* Usman Noor, "Solar-Powered Desalination—Solving the Global Water Crisis," 8MSolar, March 17, 2025, https:/8msolar.com/solar-powered-desalination-solving-the-global-water-crisis/.

31 *In the Middle East and North Africa:* "Cheap Solar Gives Desalination Its Moment in the Sun," *Financial Times*, April 28, 2024, https://www.ft.com/content/bb01b510-2c64-49d4-b819-63b1199a7f26.

31 *thirteen sextillion liters of water:* Mohammed Sanjid Thavalengal, et al., "Progress and Prospects of Air Water Harvesting System for Remote Areas: A Comprehensive Review," *Energies* 16, no. 6 (2023): 2686, https://doi.org/10.3390/en16062686.

32 *Atmospheric water capture allows us:* James Dinneen, "Hospital Hit by Hurricane Milton Gets System to Grab Water from Air," *New Scientist*, October 11, 2024, https://www.newscientist.com/article/2451657-hospital-hit-by-hurricane-milton-gets-system-to-grab-water-from-air/; Chris Stokel-Walker, "Water from the Sky: Aquaria Atmospheric Water Generators," *Time*, October 30, 2024, https://time.com/7094789/aquaria-atmospheric-water-generators/.

32 *ten billion humanoid robots:* "Elon Musk: 10 Billion Humanoid Robots by 2040 at $20K–$25K Each," Reuters, October 29, 2024, https://www.reuters.com/technology/elon-musk-10-billion-humanoid-robots-by-2040-20k-25k-each-2024-10-29/.

35 *Hollywood made the same mistake:* Allen J. Scott, "Hollywood in the Era of Globalization," YaleGlobal Online, November 29, 2002, https://archive-yaleglobal.yale.edu/content/hollywood-era-globalization.

Chapter 3: Data-Driven Optimism

41 *a logistics revolution:* "Preventing Maternal Deaths Through Faster Blood Delivery," Zipline, https://www.zipline.com/newsroom/stories/impact/preventing-maternal-deaths-through-faster-blood-delivery.

42 *Since launching in 2016:* "About Zipline," Zipline, https://www.zipline.com/about.

42 *"[This] is about one percent":* Keller Rinaudo Cliffton, "A Mini Robot—Powered by Your Phone," TED video, 5 min., 37 sec., posted February 2013, https://www.ted.com/talks/keller_rinaudo_cliffton_a_mini_robot_powered_by_your_phone/transcript.

42 *When Zipline was founded in 2014:* Briana Neuberger, "An Exploration of Commercial Unmanned Aerial Vehicles (UAVs) Through Life Cycle Assessments" (master's thesis, Rochester Institute of Technology, 2017), https://repository.rit.edu/cgi/viewcontent.cgi?article=10693&context=theses.

43 *blood, which is medicine's hardest delivery problem:* Aryn Baker, "The American Drones Saving Lives in Rwanda," *Time*, May 31, 2018, https://www.time.com/rwanda-drones-zipline/.

43 *Pre-Zipline, Rwanda:* CDC Rwanda Team, "Rwanda's National Center for Blood Transfusion Attains International Standards," Centers for Disease Control and Prevention, March 1, 2017, https://archive.cdc.gov/www_cdc_gov/globalhealth/countries/rwanda/kabeho/2017mar/ncbt.html.

43 *Zipline's first customer:* "Zipline and the Government of Rwanda Announce a New Partnership to Serve the Entire Country with Instant Logistics," *Globe-Newswire*, December 15, 2022, https://www.globenewswire.com/news-release/2022/12/15/2574639/0/en/Zipline-and-the-Government-of-Rwanda-Announce-a-New-Partnership-to-Serve-the-Entire-Country-with-Instant-Logistics.html.

43 *Zipline supplies 75 percent:* "Zipline and the Government of Rwanda."

43 *Zipline is now the planet's largest:* "Zipline Fact Sheet," Zipline, March 15, 2023, https://www.zipline.com/about/zipline-fact-sheet.

43 *A 2022* Nature *study found:* Freda Kreier, "Drones Bearing Parcels Deliver Big Carbon Savings," *Nature*, August 5, 2022, https://www.nature.com/articles/d41586-022-02101-3.

43 *But if you're looking for a statistic:* H. Harriet Jeon, Claudio Lucarelli, Jean Baptiste Mazarati, Donatien Ngabo, and Hummy Son, "Last-Mile Blood Delivery in Health Care: Drone Delivery for Blood Products in Rwanda," October 12, 2022, https://papers.ssrn.com/sol3/papers.cfm?abstract_id=4214918.

44 *Ten million—that's how many:* Max Roser, "How Many Animals Get Slaughtered Every Day?," Our World in Data, September 26, 2023, https://ourworldindata.org/how-many-animals-get-slaughtered-every-day.

44 *If the planet's population climbs:* "Global Agriculture Towards 2050," How to Feed the World 2050, High-Level Expert Forum, Rome 12–13 October 2009 (Food and Agriculture Organization of the United Nations, 2009), https://www.fao.org/fileadmin/templates/wsfs/docs/Issues_papers/HLEF2050_Global_Agriculture.pdf.

44 *one-third of the planet's arable land:* "Hannah Ritchie and Max Roser, "Half of the World's Habitable Land Is Used for Agriculture," Our World in Data, 2019, https://ourworldindata.org/global-land-for-agriculture.

44 *It takes around twelve thousand liters:* Ritchie and Roser, "Half of the World's Habitable Land Is Used for Agriculture."

44 *Add in greenhouse gas emissions:* Oliver Milman, "Meat Accounts for Nearly 60% of All Greenhouse Gases from Food Production, Study Finds," *The Guardian*, September 13, 2021, https://www.theguardian.com/environment/2021/sep/13/meat-greenhouses-gases-food-production-study.

45 *Between 2012 and 2022: The State of Food Security and Nutrition in the World 2023: Urbanization, Agrifood Systems Transformation and Healthy Diets Across the Rural–Urban Continuum. Rome* (Food and Agriculture Organization of the United Nations, 2023), https://doi.org/10.4060/cc3017en.

45 *weird NASA paper:* Lisa Lupo, "Food in Space: Defying (Micro)Gravity to Feed Our Astronauts," *Quality Assurance*, March 31, 2015, https://www.qualityassurancemag.com/article/qa0415-food-in-space-nasa/.

45 *In 2001 . . . German scientists tried:* Ian Sample, "Fish Fillets Grow in Tank," *New Scientist*, March 20, 2002, https://www.newscientist.com/article/dn2066-fish-fillets-grow-in-tank/.

45 *later be called* cultured meat*:* Sample, "Fish Fillets Grow in Tank."

45 *in Nigeria, Tetrick noticed:* Larissa Zimberoff, "Why Plant-Based Foods Pioneer Josh Tetrick Just Won't Quit," *Fast Company*, January 22, 2024, https://www.fastcompany.com/91012932/josh-tetrick-plant-based-foods-pioneer-eat-just-good-meat.

46 *He founded Eat Just:* Zimberoff, "Why Plant-Based Foods Pioneer Josh Tetrick Just Won't Quit."

46 *Since launch, they've replaced:* "Delicious Eggs Without All That Bird Flu," Eat Just, accessed September 8, 2025, https://www.ju.st/stories/bird-flu.

46 *In 2017, Eat Just blended these processes:* "Delicious Eggs."

46 *Currently, Eat Just is developing:* Elaine Watson, "Eat Just CEO: Large-Scale Cultivated Meat Production Requires a New Operating Model," AgFunderNews, March 6, 2024, https://agfundernews.com/eat-just-ceo-large-scale-cultivated-meat-production-requires-a-new-operating-model.

46 *first lab-grown burger cost $330,000:* Henry Fountain, "A Lab-Grown Burger Gets a Taste Test," *New York Times*, August 5, 2013, https://www.nytimes.com/2013/08/06/science/a-lab-grown-burger-gets-a-taste-test.html.

47 *the first holographic video display:* Jordan Harbinger, "Mary Lou Jepsen & Rob Reid | The Future of Telepathy & Affordable Healthcare," *The Jordan Harbinger Show*, June 13, 2018, https://www.jordanharbinger.com/mary-lou-jepsen-rob-reid-the-future-of-telepathy-and-affordable-healthcare/.

47 *OLPC drove that down to $180:* Adi Robertson, "OLPC's $100 Laptop Was Going to Change the World—Then It All Went Wrong," *The Verge*, April 16, 2018, https://www.theverge.com/2018/4/16/17233946/olpc-100-dollar-laptop-education-where-is-it-now.

48 *there are about 13,000 MRI machines:* "U.S. Magnetic Resonance Imaging Market Size & Outlook," Grand View Research, December 15, 2022, https://www.grandviewresearch.com/horizon/outlook/magnetic-resonance-imaging-market/united-states.

48 *there are about 440:* "Number of Magnetic Resonance Imaging (MRI) Units in Mexico in 2022, by Establishment," Statista, https://www.statista.com/statistics/1454599/number-magnetic-resonance-imaging-units-by-establishment-mexico/.

48 *According to a 2021* Lancet *study:* Kenneth A. Fleming et al., "The *Lancet* Commission on Diagnostics: Transforming Access to Diagnostics," *The Lancet* 398, no. 10315 (2021): 1997–2050, https://www.thelancet.com/journals/lancet/article/PIIS0140-6736(21)00673-5/fulltext.

48 *AI is accelerating democratization:* Naresh Kasoju et al., "Digital Health: Trends, Opportunities and Challenges in Medical Devices, Pharma and Bio-Technology,"

CSI Transactions on ICT 11 (April 2023): 11–30, https://doi.org/10.1007/s4 0012-023-00380-3.

48 *Around 2016, Jepsen decided:* Jonathan D. Grinstein, "Device Diva: Openwater Uses Cell Phone Chips to Open-Source Medical Devices," *Inside Precision Medicine*, August 27, 2024, https://www.insideprecisionmedicine.com /topics/precision-medicine/device-diva-openwater-uses-cell-phone-chips-to -open-source-medical-devices/.

49 *uses near-infrared light:* "Company Profile: Openwater," Focused Ultrasound Foundation, February 21, 2024, https://www.fusfoundation.org/posts/com pany-profile-openwater/.

49 *strokes, the number two killer:* Valery L. Feigin, "World Stroke Organization: Global Stroke Fact Sheet 2025," *International Journal of Stroke* 20, no. 2 (January 2025): 132–44, https://pmc.ncbi.nlm.nih.gov/articles/PMC11786524/.

49 *Every minute after a stroke begins:* Jeffrey L. Saver, "Time Is Brain—Quantified," *Stroke* 37, no. 1 (January 2006): 263–66, https://pubmed.ncbi.nlm.nih.gov /16339467/.

49 *If you can get a stroke patient:* J. R. Marler et al., "Early Stroke Treatment Associated with Better Outcome: The NINDS rt-PA Stroke Study," *Neurology* 55, no. 11 (December 2000), https://www.neurology.org/doi/10.1212 /WNL.55.11.1649.

49 *The disease is almost always fatal:* Fidan Şeker-Polat, Nareg Pinarbasi Degirmenci, Ihsan Solaroglu, and Tugba Bagci-Onder, "Tumor Cell Infiltration into the Brain in Glioblastoma: From Mechanisms to Clinical Perspectives," *Cancers* 14, no. 2 (January 2022): 443, https://doi.org/10.3390/cancers14020443.

50 *In preclinical trials:* Zehra E. F. Demir and Natasha D. Sheybani, "Therapeutic Ultrasound for Multimodal Cancer Treatment: A Spotlight on Breast Cancer," *Annual Review of Biomedical Engineering* 27 (2025): 371–402, https:// doi.org/10.1146/annurev-bioeng-103023-111151.

51 *In 2024, 250 million children:* UNESCO, "251M Children and Youth Still Out of School, Despite Decades of Progress," press release, November 8, 2024, https://www.unesco.org/en/articles/251m-children-and-youth-still-out -school-despite-decades-progress-unesco-report.

51 *Another 600 million:* UNESCO, "617 Million Children and Adolescents Not Getting Minimum Reading and Math," news, September 20, 2017, https:// www.unesco.org/en/articles/617-million-children-and-adolescents-not-get ting-minimum-reading-and-math.

51 *"It was a surreal conversation":* Nicholas Kristof, "Meet Sultana, the Taliban's Worst Fear," *New York Times*, June 4, 2016, https://www.nytimes.com/2016 /06/05/opinion/sunday/meet-sultana-the-talibans-worst-fear.html.

52 *In 2016, Nicholas Kristof heard:* Kristof, "Meet Sultana."

52 *when Sultana got kicked out:* Matt Lynley, "Khan Academy Snags $5 Million to Blow Up Education," *Business Insider*, November 5, 2011, https://

www.businessinsider.com/khan-academy-snags-5-million-to-blow-up-education-2011-11.

52 *In 2025, the figure was over 150 million users:* "Annual Report: SY24–25," Khan Academy, June 30, 2025, https://annualreport.khanacademy.org.

52 *According to a 2024 study:* Phil Grimaldi, "Multiple Studies Show Khan Academy Drives Learning Gains: Evidence for Our Platform's Effectiveness," Khan Academy, November 16, 2023, https://blog.khanacademy.org/multiple-studies-show-khan-academy-drives-learning-gains-evidence-for-our-platforms-effectiveness/.

52 *Khan believes that students:* Bogdan Yamkovenko, Kodi Weatherholtz, and Phil Grimaldi, "New Study Finds Every Minute Spent on Khan Academy Can Lead to Learning Gains," Khan Academy, September 28, 2023, https://blog.khanacademy.org/new-study-finds-every-minute-spent-on-khan-academy-can-lead-to-learning-gains/.

53 *Rice is another major component:* "Rice Sector at a Glance," Economic Research Service, United States Department of Agriculture, updated January 7, 2025, https://www.ers.usda.gov/topics/crops/rice/rice-sector-at-a-glance.

53 *For half a century, researchers have tried:* Shilai Zhang et al., "Sustained Productivity and Agronomic Potential of Perennial Rice," *Nature Sustainability* 6 (2023): 28–38, https://www.nature.com/articles/s41893-022-00997-3.

53 *Rice's annual planting cycles deplete:* Tran Dang Xuan, Tran Thi Ngoc Minh, Ramin Rayee, Ngo Duy Dong, and Nguyen Xuan, "Advances in Mitigating Methane Emissions from Rice Cultivation: Past, Present, and Future Strategies," *Environmental Science and Pollution Research* 32 (2025): 20232–47, https://link.springer.com/article/10.1007/s11356-025-36776-8.

54 *home to 1.42 billion people:* "China's Population Falls for a Third Consecutive Year," Reuters, January 17, 2025, https://www.reuters.com/world/china/chinas-population-falls-third-consecutive-year-2025-01-17/.

54 *In early 2023, scientists:* Nectar Gan, "China Says It Successfully Cloned 3 Highly Productive 'Super Cows,'" CNN, February 1, 2023, https://www.cnn.com/2023/02/01/business/china-super-cow-cloning-intl-hnk-scn/index.html.

54 *Adoption rates for AI-powered:* Ridha Guebsi, Sonia Mami, and Karem Chokmani, "Drones in Precision Agriculture: A Comprehensive Review of Applications, Technologies, and Challenges," *Drones* 8, no. 11 (2024): 686, https://doi.org/10.3390/drones8110686.

54 *In this same period, vertical farming:* "Earth Day with AeroFarms: A Look at Vertical Farming's Environmental Impact!," AeroFarms, April 19, 2024, https://www.aerofarms.com/earth-day-with-aerofarms-a-look-at-vertical-farmings-environmental-impact/.

55 *A decade later, autonomous taxis operate:* "Autonomous Vehicles: Timeline and Roadmap Ahead," World Economic Forum, white paper, April 2025, https://reports.weforum.org/docs/WEF_Autonomous_Vehicles_2025.pdf.

55 *Tesla's Waymo competition, Cybercab:* Eric Stafford and Elana Scherr, "Tesla Robotaxi Is a Driverless Car That Will Cost Under $30,000," *Car and Driver*, October 10, 2024, https://www.caranddriver.com/news/a62567491/tesla-robotaxi-reveal/.

55 *Morgan Stanley estimates:* Adam Spatacco, "The Low Altitude Market Could Be Worth $9 Trillion by 2050, According to Morgan Stanley. This Cathie Wood Stock Is My Top Pick to Dominate the Opportunity (Hint: It's Not Archer Aviation)," *Motley Fool*, July 12, 2025, https://www.fool.com/investing/2025/07/12/the-low-altitude-market-could-be-worth-9-trillion/.

55 *CRISPR gene-editing therapies:* "FDA-Approved Gene Therapies," Boston Children's Hospital, August 31, 2022, https://www.childrenshospital.org/programs/gene-therapy-program/fda-approved-gene-therapies.

55 *BlueRock Therapeutics, for instance:* "BlueRock Therapeutics Announces Publication in *Nature* of 18-Month Data from Phase 1 Clinical Trial for Bemdaneprocel, an Investigational Cell Therapy for Parkinson's Disease," BlueRock Therapeutics, April 16, 2025, https://www.bluerocktx.com/bluerock-therapeutics-announces-publication-in-nature-of-18-month-data-from-phase-1-clinical-trial-for-bemdaneprocel-an-investigational-cell-therapy-for-parkinsons-disease/.

55 *more than 250 million children:* Victoria Masterson, "This School-in-a-Bag Helps Educate the Most Remote Areas," World Economic Forum, July 14, 2021, https://www.weforum.org/stories/2021/07/tespack-solar-bag-access-school/.

56 *Coursera alone serves 125 million learners:* "About Coursera," Coursera, https://about.coursera.org/press#:~:text=insights%20about%20Coursera.

56 *AI-tutoring systems like Squirrel AI:* Jyothi Kasinath, "How Technology Helped BYJU'S Scale Up to 100 M Users," Salesforce India Blog, Salesforce, August 26, 2021, https://www.salesforce.com/in/blog/how-technology-helped-byjus-scale-up-to-100-million-users-without-sacrificing-quality/; Russell Flannery, "Derek Li and Squirrel Ai Aim to Lead the Future of AI-Driven Education," *Forbes*, February 18, 2025, https://www.forbes.com/sites/forbeschina/2025/02/18/derek-li-and-squirrel-ai-aim-to-lead-the-future-of-ai-driven-education/.

56 *Even in regions without stable internet:* "Kolibri," HundrED, June 2022, https://hundred.org/en/innovations/kolibri.

56 *And all of this helps explain:* "Global Adult Literacy Rate Aged 15 Years and Older from 1976 to 2023 by Gender," Statista, https://www.statista.com/statistics/997360/global-adult-and-youth-literacy/.

PART 2

Chapter 4: One Billion Times Smarter

60 *In 2022, McKinsey estimated:* "The Economic Potential of Generative AI: The Next Productivity Frontier," McKinsey, June 14, 2023, https://www.mckinsey.com/capabilities/mckinsey-digital/our-insights/the-economic-potential-of-generative-ai-the-next-productivity-frontier.

60 *PricewaterhouseCoopers projects:* PricewaterhouseCoopers, "Sizing the Prize: What's the Real Value of AI for Your Business and How Can You Capitalise?," 2017, https://www.pwc.com/gx/en/issues/analytics/assets/pwc-ai-analysis-sizing-the-prize-report.pdf.

60 *In healthcare, AI has improved diagnostics:* Scott Mayer McKinney et al., "International Evaluation of an AI System for Breast Cancer Screening," *Nature* 577 (2020): 89–94, https://doi.org/10.1038/s41586-019-1799-6.

60 *The US Department of Energy found:* Keith J. Benes, Joshua E. Porterfield, and Charles Yang, "AI for Energy: Opportunities for a Modern Grid and Clean Energy Economy," Department of Energy, April 2024, https://www.energy.gov/sites/default/files/2024-04/AI%20EO%20Report%20Section%205.2g%28i%29_043024.pdf.

60 *In farming, AI companies:* A. S. Yeshe, P. H. Gourkhede, and P. H. Vaidya, "Blue River Technology: Futuristic Approach of Precision Farming," *Just Agriculture* 2, no. 7 (March 2022), https://justagriculture.in/files/newsletter/2022/march/04.pdf.

61 *With funding from the Rockefeller Foundation:* J. McCarthy, M. L. Minsky, N. Rochester, and C. E. Shannon, "A Proposal for the Dartmouth Summer Research Project on Artificial Intelligence," Stanford University, August 31, 1955, https://www-formal.stanford.edu/jmc/history/dartmouth/dartmouth.html.

61 *They didn't solve it:* McCarthy, Minsky, Rochester, and Shannon, "A Proposal."

62 *In the late 1950s, after the close:* "Newell, Simon & Shaw Develop the First Artificial Intelligence Program, 1955 to 7/1956," Jeremy Norman's History ofInformation.com, https://www.historyofinformation.com/detail.php?id=742.

63 *MYCIN started to diagnose:* Sergio Sancho Azcoitia, "MYCIN: The Beginning of Artificial Intelligence in Medicine," *Telefónica Tech*, November 15, 2018, https://telefonicatech.com/en/blog/mycin-the-beginning-of-artificial-intelligence-in-medicine.

63 *while DENDRAL cracked molecular:* "DENDRAL—The First 'Expert System,'" Boston Global Forum, April 6, 2025, https://bostonglobalforum.org/news/dendral-the-first-expert-system/.

63 *AI clawed its way back:* Jay McClelland, "From Brain to Machine: The Unexpected Journey of Neural Networks," Stanford University Human-Centered Artificial Intelligence, November 18, 2024, https://hai.stanford.edu/news/from-brain-to-machine-the-unexpected-journey-of-neural-networks.

63 *In 1997, IBM's Deep Blue defeated:* "Deep Blue," IBM, https://www.ibm.com/history/deep-blue.

64 *In the early 1980s DARPA began experimenting:* "Driving Forces: Autonomous Land Vehicles," Lockheed Martin, October 1, 2020, https://www.lockheedmartin.com/en-us/news/features/history/alv.html.

64 *It looked like a four-page paper:* G. E. Hinton and R. R. Salakhutdinov, "Reducing the Dimensionality of Data with Neural Networks," *Science* 313, no. 5786 (2006): 504–7, https://doi.org/10.1126/science.1127647.

65 *Introducing the concept of "deep belief networks":* Geoffrey E. Hinton, Simon Osindero, and Yee-Whye Teh, "A Fast Learning Algorithm for Deep Belief Nets," *Neural Computation* 18, no. 7 (2006): 1527–54, https://doi.org/10.1162/neco.2006.18.7.1527, https://www.cs.toronto.edu/~hinton/absps/fastnc.pdf.

65 *In 2012, Alex Krizhevsky:* Alex Krizhevsky, Ilya Sutskever, and Geoffrey E. Hinton, "ImageNet Classification with Deep Convolutional Neural Networks," *Communications of the ACM* 60 (2012): 84–90, https://doi.org/10.1145/3065386.

65 *AlexNet slashed the error rate:* Krizhevsky, Sutskever, and Hinton, "ImageNet Classification with Deep Convolutional Neural Networks."

65 *"This was the big moment":* Lloyd Lee, "Nvidia CEO Jensen Huang Shouts Out OpenAI Cofounder Ilya Sutskever for Sparking 'the Big Bang of Deep Learning,'" *Business Insider*, June 22, 2024, https://www.businessinsider.com/nvidia-ceo-jensen-huang-openai-ilya-sutskever-ai-revolution-2024-6.

66 *The six years from 2006 to 2012:* Giuliano Giacaglia, "A Brief Overview of Deep Learning," Holloway, November 2, 2022, https://www.holloway.com/g/making-things-think/sections/a-brief-overview-of-deep-learning.

66 *In recognition of this work:* "Fathers of Deep Learning Revolution Receive ACM A. M. Turing Award," Association for Computing Machinery, 2018 ACM A. M. Turing Award Laureates, https://awards.acm.org/about/2018-turing.

66 *Then, in 2024:* Rahul Kalvapalle, "Geoffrey Hinton Wins Nobel Prize in Physics," University of Toronto, October 8, 2024, https://www.utoronto.ca/news/geoffrey-hinton-wins-nobel-prize.

66 *For all these reasons:* Jason Ma, "'Godfather of AI' Says the Technology Will Create Massive Unemployment and Send Profits Soaring—That Is the Capitalist System,'" *Fortune*, September 6, 2025, https://fortune.com/2025/09/06/godfather-of-ai-geoffrey-hinton-massive-unemployment-soaring-profits-capitalist-system/.

67 *"Attention Is All You Need":* Ashish Vaswani et al., "Attention Is All You Need," *Advances in Neural Information Processing Systems* (NIPS 2017) 30 (2017): 5998–6008, https://papers.nips.cc/paper/7181-attention-is-all-you-need.pdf.

67 *After a year in the trenches:* Alec Radford, Karthik Narasimhan, Tim Salimans, and Ilya Sutskever, "Improving Language Understanding by Generative Pre-Training," OpenAI, June 11, 2018, https://cdn.openai.com/research-covers/language-unsupervised/language_understanding_paper.pdf.

67 *Less than a year later, GPT-2 arrived:* Alec Radford, Jeffrey Wu, Rewon Child, David Luan, Dario Amodei, and Ilya Sutskever, "Language Models Are Unsupervised Multitask Learners," OpenAI, 2019, https://cdn.openai.com/better-language-models/language_models_are_unsupervised_multitask_learners.pdf.

67 *The world changed on November 30, 2022:* "Better Language Models and Their Implications," OpenAI, June 11, 2020, https://openai.com/index/better-language-models/.

68 *In five days, the system:* Fabio Duarte, "Number of ChatGPT Users (October 2025)," Exploding Topics, October 2, 2025, https://explodingtopics.com/blog/chatgpt-users.

68 *AI enthusiasts invested over $200 billion:* Anton Shilov, "Meta to Build 2GW Data Center with Over 1.3 Million Nvidia AI GPUs—Invest $65 Billion in AI in 2025," Tom's Hardware, January 25, 2025, accessed September 9, 2025, https://www.tomshardware.com/tech-industry/artificial-intelligence/meta-to-build-2gw-data-center-with-over-1-3-million-nvidia-ai-gpus-invest-usd65b-in-ai-in-2025.

70 *In April 2024, Aschenbrenner was dismissed:* Ana Altchek, "Ex–OpenAI Researcher Speaks Out About Why He Was Fired: 'I Ruffled Some Feathers,'" *Business Insider*, June 5, 2024, https://www.businessinsider.com/former-openai-researcher-leopold-aschenbrenner-interview-firing-2024-6.

70 *Aschenbrenner published a lengthy white paper:* Leopold Aschenbrenner, "Situational Awareness: The Decade Ahead," June 2024, https://situational-awareness.ai/wp-content/uploads/2024/06/situationalawareness.pdf.

71 *when, exactly, we will reach AGI:* Aschenbrenner, "Situational Awareness."

71 *In 1964, less than a decade after:* Arthur C. Clarke, "1964: Arthur C. Clarke Predicts the Future," *Horizon*, BBC Archive, originally broadcast September 21, 1964, https://blog.biocomm.ai/2025/01/16/1964-arthur-c-clarke-predicts-the-future-bbc-archive/.

72 *What Gawdat tells us, in his book:* Mo Gawdat, *Scary Smart: The Future of Artificial Intelligence and How You Can Save Our World* (Bluebird, 2021).

72 *"Be nice":* Gawdat, *Scary Smart.*

72 *"Instead of containing them":* Gawdat, *Scary Smart.*

72 *Oxford philosopher Nick Bostrom:* Nick Bostrom, *Superintelligence: Paths, Dangers, Strategies* (Oxford University Press, 2014).

72 *For similar reasons, University of California:* Stuart Russell, *Human Compatible: Artificial Intelligence and the Problem of Control* (Viking, 2019).

73 *In early 2024, when Anthropic released:* "Claude-3 Surpasses Human Average IQ, Anthropic Leads AI Intelligence into a New Era," AI Daily, AIbase, April 22, 2025, https://www.aibase.com/news/17403.

74 *in 2009, leaving academia:* Rob Toews, "8 Leading Women in the Field of AI," *Forbes*, December 13, 2020, https://www.forbes.com/sites/robtoews/2020/12/13/8-leading-women-in-the-field-of-ai/#.

74 *Affectiva has worked with 28 percent:* "Affectiva's Global Emotion AI Advertising Database," Affectiva, October 27, 2020, https://www.affectiva.com/news-item/research-shows-advertising-is-having-a-more-emotional-impact-on-consumers-but-not-all-advertisers-are-using-emotions-effectively/.

74 *In 2012, she was named:* "Alumna Rana el Kaliouby AUC's Newest, Youngest Trustee," American University in Cairo, October 13, 2015, https://www.aucegypt.edu/news/stories/alumna-rana-el-kaliouby-aucs-newest-youngest-trustee.

74 *A few years later, she was named again:* "Dr. Rana el Kaliouby," XPRIZE, https://www.xprize.org/about/people/dr-rana-el-kaliouby.

74 *Her memoir,* Girl Decoded: Rana el Kaliouby, *Girl Decoded: A Scientist's Quest to Reclaim Our Humanity by Bringing Emotional Intelligence to Technology* (Currency, 2020).

75 *As Bill Gates wrote in his blog:* Bill Gates, "AI is about to completely change how you use computers. And upend the software industry." GatesNotes.com, November 9, 2023, https://www.gatesnotes.com/ai-agents.

76 *In early 2025, Salesforce made headlines:* "Salesforce CEO Confirms AI Now Handles 30%–50% of Company's Work, Drives Major Workforce and Productivity Shift," MLQ.ai, June 30, 2025, https://mlq.ai/news/salesforce-ceo-confirms-ai-now-handles-30-50-of-companys-work-drives-major-workforce-and-productivity-shift/.

76 *In June 2025, OpenAI and Mattel:* "Mattel and OpenAI Announce Strategic Collaboration," Mattel, June 12, 2025, https://corporate.mattel.com/news/mattel-and-openai-announce-strategic-collaboration.

Chapter 5: Surfing the Tsunami

80 *In the United States alone:* Arthur Caplan and Brendan Parent, "Organ Transplantation," Hastings Center for Bioethics, June 7, 2022, https://www.thehastingscenter.org/briefingbook/organ-transplantation/.

80 *Globally, millions suffer:* Caplan and Parent. "Organ Transplantation."

81 *Not long ago, surgeons:* Jen Middleton, "What Is the Time Frame for Transplanting Organs?," *Donor Essentials* (blog), Donor Alliance, February 21,

2025, https://www.donoralliance.org/newsroom/donation-essentials/what-is-the-time-frame-for-transplanting-organs/.

81 *Yet Kamen has fifteen hundred patents:* "Patents by Inventor Dean Kamen," Justia Patents, https://patents.justia.com/inventor/dean-kamen.

81 *Meanwhile, the FIRST Robotics Competition:* "About FIRST: Our Mission, Purpose & Values," FIRST, November 7, 2023, https://www.firstinspires.org/about.

84 *Forty million Roombas:* "History," iRobot, https://about.irobot.com/history.

84 *The Roomba's "body plan":* Tyler Greenawalt, "Amazon Has More Than 1 Million Robots That Sort, Lift, and Carry Packages—See Them in Action," Amazon, October 22, 2025, https://www.aboutamazon.com/news/operations/amazon-robotics-robots-fulfillment-center.

85 *In 2013, they introduced Atlas:* Will Knight, "Meet Atlas, the Robot Designed to Save the Day," *MIT Technology Review*, July 12, 2013, https://www.technologyreview.com/2013/07/12/15691/meet-atlas-the-robot-designed-to-save-the-day/.

85 *Optimus—now slated:* Hallie Steiner, "Elon Musk Reveals Massive Plans for Tesla and Optimus—'Things Are Really Going to Go Ballistic Next Year,'" *Fortune*, January 30, 2025, https://fortune.com/2025/01/30/elon-musk-reveals-massive-plans-tesla-optimus-self-driving-cars-humanoid-robots/.

85 *Amazon and Agility Robotics:* Truman Lewis, "Report Says Amazon Developing Humanoid Delivery Robots," Consumer Affairs, June 5, 2025, https://www.consumeraffairs.com/news/report-says-amazon-developing-humanoid-delivery-robots-060525.html.

85 *Meanwhile, by mid-2026:* "OpenAI-Backed Startup to Begin in-Home Tests of Humanoid Robot, Neo Gamma," *Indian Express*, March 24, 2025, https://indianexpress.com/article/technology/tech-news-technology/openai-startup-1x-in-home-tests-humanoid-robot-neo-gamma-9901369/.

85 *Since China also dominates:* Caiwei Chen, "China's EV Giants Are Betting Big on Humanoid Robots," *MIT Technology Review*, February 14, 2025, https://www.technologyreview.com/2025/02/14/1111920/chinas-electric-vehicle-giants-pivot-humanoid-robots/.

85 *The Robot+ initiative:* "Will the United States or China Lead in Humanoid Robotics?," *Special Competitive Studies Project*, October 15, 2024, https://scsp222.substack.com/p/will-the-united-states-or-china-lead.

85 *Advances in AI, actuator technology:* Bao Tran, "The Impact of AI on Battery Technology: How AI Is Improving Battery Performance (Key Stats)," PatentPC, October 9, 2025, https://patentpc.com/blog/the-impact-of-ai-on-battery-technology-how-ai-is-improving-battery-performance-key-stats.

86 *Both Figure AI and Tesla:* Cyrus Cole, "Tesla's AI and Robotics Pivot: A High-Stakes Gamble for Long-Term Investors?," AInvest, September 4, 2025,

https://www.ainvest.com/news/tesla-ai-robotics-pivot-high-stakes-gamble-long-term-investors-2509/.

86 *Nearly half a million autonomous:* Scott Dresser, "Amazon Launches a New AI Foundation Model to Power Its Robotic Fleet and Deploys Its 1 Millionth Robot," Amazon, June 30, 2025, https://www.aboutamazon.com/news/operations/amazon-million-robots-ai-foundation-model.

86 *Amazon reports that robots improve:* Dresser, "Amazon Launches a New AI Foundation Model."

86 *These gains are mirrored:* Rajarshi Chatterjee, "How Drones Are Revolutionising Air Cargo Sustainability," *Stat Trade Times,* May 31, 2025, https://www.stattimes.com/drones/how-drones-are-revolutionising-air-cargo-sustainability-1355444.

86 *By 2030, robots are anticipated:* "Surgical Robots Market (2025–2030)," Market Analysis Report, Grand View Research, December 31, 2023, accessed September 11, 2025, https://www.grandviewresearch.com/industry-analysis/surgical-robot-market.

86 *Improvements in motor accuracy:* Jack Ng Kok Wah, "The Rise of Robotics and AI-Assisted Surgery in Modern Healthcare," *Journal of Robotic Surgery* 19, no. 1 (2025): 311, https://pubmed.ncbi.nlm.nih.gov/40540146/.

86 *Robots in eldercare are projected:* "Japan: Demographic Changes," Economic and Social Commission for Asia and the Pacific, https://www.population-trends-asiapacific.org/data/population-ageing/JPN.

86 *robots capable of precision farming:* Amit Sharma, et al., "Artificial Intelligence and Internet of Things Oriented Sustainable Precision Farming: Towards Modern Agriculture," *Open Life Sciences* 18, no. 1 (2023): 20220713, https://doi.org/10.1515/biol-2022-0713.

86 *This efficiency will be essential:* Janet Ranganathan, Richard Waite, Tim Searchinger, and Craig Hanson, "How to Sustainably Feed 10 Billion People by 2050, in 21 Charts," World Resources Institute, December 5, 2018, https://www.wri.org/insights/how-sustainably-feed-10-billion-people-2050-21-charts.

89 *Ben Lamm, CEO of Colossal Biosciences:* Ben Swanger, "Colossal Becomes First Private Startup Founded in Texas to Earn $10 Billion Valuation," *D,* January 15, 2025, https://www.dmagazine.com/business-economy/2025/01/colossal-earns-10-billion-valuation/#:~:text=Colossal%20Becomes%20First%20Private%20Startup,it%20has%20trillion%2Ddollar%20potential.&text=Dallas%2Dbased%20Colossal%20has%20put,class%20of%20businesses%3A%20a%20decacorn.

89 *Three years later:* John McKenna, "This Animal Went Extinct Twice," World Economic Forum, December 11, 2017, https://www.weforum.org/stories/2017/12/bucardo-goat-went-extinct-twice/.

89 *By 2020, CRISPR-Cas9:* Sameh A. Abdelnour, Long Xie, Abdallah A. Hassanin, Erwei Zuo, and Yangqing Lu, "The Potential of CRISPR/Cas9 Gene Editing as a Treatment Strategy for Inherited Diseases," *Frontiers in Cell and Developmental Biology* 9 (December 2021), https://doi.org/10.3389/fcell.2021.699597.

89 *Harvard geneticist George Church:* Stephanie Dutchen, "A Mammoth Solution: Scientists Look to Extinct Genes to Protect Endangered Species, Climate," Harvard Medical School, November 12, 2021, https://hms.harvard.edu/news/mammoth-solution.

90 *It's been four thousand years:* Carl Zimmer, "The Last Stand of the Woolly Mammoths," *New York Times*, July 27, 2024, https://www.nytimes.com/2024/06/27/science/mammoth-genes-wrangel.html.

90 *It's been more than three hundred years:* "The Dodo Bird: The Real Facts About This Icon of Extinction," Natural History Museum, https://www.nhm.ac.uk/discover/the-dodo-bird-the-real-facts-about-this-icon-of-extinction.html.

91 *Colossal has eight- and nine-figure deals:* "Our Ancient DNA Research: A Kaleidoscope of Ancient Life," Species: De-extinction for the Survival of Our Planet, Colossal Laboratories & Biosciences, January 31, 2023, https://colossal.com/species/.

91 *In April 2025, Lamm announced:* Jeffrey Kluger, "The Return of the Dire Wolf," *Time*, April 7, 2025, https://time.com/7274542/colossal-dire-wolf/.

91 *All that was left behind:* Kluger, "The Return of the Dire Wolf."

91 *Colossal's scientists compared:* Katie Hunt, "Scientists Say They Have Resurrected the Dire Wolf," CNN, April 8, 2025, https://www.cnn.com/2025/04/07/science/dire-wolf-de-extinction-cloning-colossal.

91 *Next, they edited fourteen genes:* Jeffrey Kluger, "The Science Behind the Return of the Dire Wolf," *Time*, April 7, 2025, https://time.com/7275439/science-behind-dire-wolf-return/.

92 *Finally, the edited nuclei:* Kluger, "The Science Behind the Return of the Dire Wolf."

93 *Meanwhile, NASA was building satellites:* Shlomo Sprung, "The Most Expensive Failed Space Missions of All Time," *Business Insider*, accessed September 11, 2025, https://www.businessinsider.com/the-most-expensive-failed-space-missions-of-all-time-2012-8.

94 *Today, Planet has some five hundred:* "Our Next-Generation Satellite Constellation Pelican Is Expected to Deliver Very-High-Resolution and Rapid-Revisit Capabilities," Planet Pulse, Planet, April 21, 2022, https://www.planet.com/pulse/next-generation-satellite-pelican/.

94 *The fleet represents a seismic shift:* Nikolai Khlystov and Gayle Markovitz, "Space Is Booming. Here's How to Embrace the $1.8 Trillion Opportunity," World Economic Forum, April 8, 2024, https://www.weforum.org/stories/2024/04/space-economy-technology-invest-rocket-opportunity/.

96 *Six hundred million people:* Brad Plumer, "There Are 620 Million People in Africa Without Electricity. Here's Where They Live," *Vox*, October 13, 2014, https://www.vox.com/2014/10/13/6970513/africa-electricity-620-million-people-map-gas-coal-solar.

96 *Yet Africa receives over sixty million:* Mogomotsi Magome, "Africa's Solar Energy Potential Makes for a Bright Future for Renewable Power," Associated Press, September 7, 2025, https://apnews.com/article/africa-solar-tower-renewable-energy-698216c8520dbaf9385e7d41515901b1.

96 *In some regions:* Abdullahi Mohamed Samatar et al., "The Utilization and Potential of Solar Energy in Somalia: Current State and Prospects," *Energy Strategy Reviews* 48 (July 2023): 101108, https://doi.org/10.1016/j.esr.2023.101108.

97 *An eighty-five kilowatt mini-grid:* Amy Yee, "Solar Mini-Grids Give Nigeria a Power Boost," *New York Times*, December 2, 2018, https://www.nytimes.com/2018/12/02/climate/solar-mini-grids-give-nigeria-a-power-boost.html.

97 *And this metamorphosis didn't rely:* "Nigeria Suffers Power Outage After Grid Failure, Power Companies Say," Reuters, December 11, 2024, https://www.reuters.com/world/africa/nigeria-suffers-power-outage-after-grid-failure-power-companies-say-2024-12-11/.

97 *Mini-grids are a potent solution:* "Solar Mini Grids Could Power Half a Billion People by 2030—If Action Is Taken Now," World Bank, September 27, 2022, https://www.worldbank.org/en/news/press-release/2022/09/27/solar-mini-grids-could-power-half-a-billion-people-by-2030-if-action-is-taken-now.

97 *As solar mini-grids also reduce:* "Net Zero by 2050: A Roadmap for the Global Energy Sector," International Energy Agency, October 2021, https://www.energy.gov/sites/default/files/2021-12/IEA,%20Net%20Zero%20by%202050.pdf.

97 *Because of her work:* "Damilola Ogunbiyi," U.S. International Development Finance Corporation, https://www.dfc.gov/who-we-are/development-accountability-council-dac/damilola-ogunbiyi.

98 *International Renewable Energy Agency:* "Increasing World's Share of Renewable Energy Would Boost Global GDP up to $1.3 Trillion," United Nations Climate Change, January 16, 2016, https://unfccc.int/news/increasing-world-s-share-of-renewable-energy-would-boost-global-gdp-up-to-13-trillion.

98 *By 2030, solar mini-grids:* "Solar Mini Grids Could Sustainably Power 380 Million People in Africa by 2030—If Action Is Taken Now," World Bank, February 26, 2023, https://www.worldbank.org/en/news/press-release/2023/02/26/solar-mini-grids-could-sustainably-power-380-million-people-in-afe-africa-by-2030-if-action-is-taken-now.

98 *Lithium-ion battery prices:* Hannah Ritchie, "The Price of Batteries Has Declined by 97% in the Last Three Decades," Our World in Data, June 4, 2021, https://ourworldindata.org/battery-price-decline.

Chapter 6: The Dark Side of Abundance

101 *Let's take a walk through:* Dan Szczesny, "The Great Horse Manure Crisis of 1894," *Day by Day*, January 2, 2024, https://danszczesny.substack.com/p/the-great-horse-manure-crisis-of.

102 *Nationally, the horse population peaked:* Miles C. Collier, "Been Theres, Done That: When the Automobile Was the World's Technological Savior," Revs Institute, April 29, 2021, https://automedia.revsinstitute.org/been-there-done-that-when-the-automobile-was-the-worlds-technological-savior.

102 *In major cities:* Brad Smith and Carol Ann Browne, "The Day the Horse Lost Its Job," *Today in Tech* (blog), Microsoft, December 21, 2017, https://blogs.microsoft.com/today-in-tech/day-horse-lost-job/.

102 *Try forty days:* Lea C. de Hesselle and Christian Montag, "Effects of a 14-Day Social Media Abstinence on Mental Health and Well-Being: Results from an Experimental Study," *BMC Psychology* 12, 141 (March 2024), https://doi.org/10.1186/s40359-024-01611-1.

103 *Or, as biologist E. O. Wilson:* Konrad, "Timeless Minds in a Technological Age: The Enduring Legacy of E. O. Wilson," *Medium*, April 14, 2025, https://medium.com/@im_jokingcom/timeless-minds-in-a-technological-age-the-enduring-legacy-of-e-o-wilson-a97da7b4de89.

104 *Today, the Jakobshavn Glacier:* "On Thin Ice," UNH Today, University of New Hampshire, December 22, 2007, https://www.unh.edu/unhtoday/2007/12/thin-ice.

104 *In the past two decades:* Francie Diep, "U.S. Has Depleted Two Lake Eries' Worth of Groundwater Since 1900," *Popular Science*, May 21, 2013, https://www.popsci.com/environment/article/2013-05/us-depleted-two-lake-eries-worth-underground-water-1900-study-finds/.

104 *Since the Industrial Revolution:* "Keep the 1.5°C Goal Alive, Experts and Civil Society Urge on 'Energy Day' at COP27," UN News, November 15, 2022, https://news.un.org/en/story/2022/11/1130622.

104 *The atmosphere now contains:* "CO2 Levels in Atmosphere Are at Their Highest in 800,000 Years," World Economic Forum, May 9, 2018, https://www.weforum.org/stories/2018/05/earth-just-hit-a-terrifying-milestone-for-the-first-time-in-more-than-800-000-years/#:~:text=%22As%20a%20scientist%2C%20what%20concerns,to%20breathe%20air%20like%20this.

105 *In 2023, wildfire smoke:* Laura Goggin, "That Day NYC Looked like Blade Runner 2049," Laura Goggin Photography, https://www.gogginphotography.com/2023/06/that-day-nyc-looked-like-blade-runner.html.

105 *In 2025, a large chunk:* Tony Briscoe and Ian James, "The L.A. Wildfires Left Neighborhoods Choking in Ash and Toxic Air. Residents Demand Answers,"

Los Angeles Times, January 30, 2025, https://www.latimes.com/environment/story/2025-01-30/la-wildfires-hazardous-waste-cleanup.

105 *In the United States alone:* "XPRIZE Launches Competition to End Destructive Wildfires," XPRIZE, April 21, 2023, https://www.xprize.org/prizes/wildfire/articles/xprize-launches-competition-to-end-destructive-wildfires.

105 *The Reno-based start-up:* Charlotte Edmond, "These Drones Can Plant 100,000 Trees a Day," World Economic Forum, June 29, 2017, https://www.weforum.org/stories/2017/06/drones-plant-100000-trees-a-day/.

106 *Bill Gates predicts:* Catherine Clifford, "Bill Gates Says Climate Tech Will Produce 8 to 10 Teslas, a Google, an Amazon and a Microsoft," CNBC, October 20, 2021, https://www.cnbc.com/2021/10/20/bill-gates-expects-8-to-10-teslas-and-a-google-amazon-and-microsoft.html.

106 *If temperatures rise another 0.9 degrees:* Ove Hoegh-Guldberg et al., "Impacts of 1.5°C of Global Warming on Natural and Human Systems," in *Global Warming of 1.5°C. An IPCC Special Report on the Impacts of Global Warming of 1.5°C Above Pre-industrial Levels and Related Global Greenhouse Gas Emission Pathways, in the Context of Strengthening the Global Response to the Threat of Climate Change, Sustainable Development, and Efforts to Eradicate Poverty* ed. V. Masson-Delmotte et al. (Cambridge University Press, 2018), 175–312, https://doi.org/10.1017/9781009157940.005, https://www.ipcc.ch/site/assets/uploads/sites/2/2022/06/SR15_Chapter_3_LR.pdf.

106 *By 2050, hundreds of millions:* "The Climate Crisis—A Race We Can Win," United Nations, https://www.un.org/en/un75/climate-crisis-race-we-can-win.

106 *In 2022, Dutch researchers:* Damian Carrington, "Microplastics Found in Human Blood for First Time," *The Guardian*, March 24, 2022, https://www.theguardian.com/environment/2022/mar/24/microplastics-found-in-human-blood-for-first-time.

106 *A few months later:* Sarah Kuta, "The Human Brain May Contain as Much as a Spoon's Worth of Microplastics, New Research Suggests," *Smithsonian*, February 4, 2025, https://www.smithsonianmag.com/smart-news/the-human-brain-may-contain-as-much-as-a-spoons-worth-of-microplastics-new-research-suggests-180985995/.

106 *Every minute, a garbage truck's worth:* James Pennington, "Every Minute, One Garbage Truck of Plastic Is Dumped into Our Oceans. This Has to Stop," World Economic Forum, October 27, 2016, https://www.weforum.org/stories/2016/10/every-minute-one-garbage-truck-of-plastic-is-dumped-into-our-oceans/.

107 *By 2050, the weight of all:* "Will There Be More Plastic than Fish in the Sea?," World Wildlife Fund, https://www.wwf.org.uk/myfootprint/challenges/will-there-be-more-plastic-fish-sea.

107 *First developed by 3M and DuPont:* Sara Samora, "The History of PFAS: From World War II to Your Teflon Pan," Manufacturing Dive, December 6,

2023, https://www.manufacturingdive.com/news/the-history-behind-forever-chemicals-pfas-3m-dupont-pfte-pfoa-pfos/698254/.

107 *According to the CDC:* "Fast Facts: PFAS in the U.S. Population," Agency for Toxic Substances and Disease Registry, November 12, 2024, https://www.atsdr.cdc.gov/pfas/data-research/facts-stats/index.html.

107 *It's estimated that we consume:* "A Breath of Fresh . . . Plastic? Humans Inhale a Credit Card's Worth of Microplastics Every Week," Plastic Pollution Coalition, March 7, 2024, https://www.plasticpollutioncoalition.org/blog/2024/3/7/humans-inhale-a-credit-cards-worth-of-microplastics-every-week.

107 *They've been detected at the summit:* Damian Carrington, "Microplastic Pollution Found Near Summit of Mount Everest," *The Guardian*, November 20, 2020, https://www.theguardian.com/environment/2020/nov/20/microplastic-pollution-found-near-summit-of-mount-everest.

107 *Major companies:* Tom Perkins, "Investors Pressure Top Firms to Halt Production of Toxic 'Forever Chemicals,'" *The Guardian*, January 6, 2023, https://www.theguardian.com/environment/2023/jan/06/pfas-toxic-forever-chemicals-manufacturers.

108 *We produce about four hundred million:* "The World's Plastic Pollution Crisis, Explained," *National Geographic*, May 28, 2025, https://www.nationalgeographic.com/environment/article/plastic-pollution.

109 *Today, 38 percent of the world's population:* Denis Campbell, "More Than Half of Humans on Track to Be Overweight or Obese by 2035—Report," *The Guardian*, March 2, 2023, https://www.theguardian.com/society/2023/mar/02/more-than-half-of-humans-on-track-to-be-overweight-or-obese-by-2035-report.

109 *The global cost of obesity:* "Economic Impact of Overweight and Obesity to Surpass $4 Trillion by 2035," World Obesity, https://www.worldobesity.org/news/economic-impact-of-overweight-and-obesity-to-surpass-4-trillion-by-2035.

109 *On the medical front, GLP-1 agonists:* David Brennan, "Pros and Cons of GLP-1 Agonists for Weight Loss," Mayo Clinic, August 14, 2025, https://communityhealth.mayoclinic.org/featured-stories/weight-loss-drugs.

109 *According to the UN, 735 million:* "The State of Food Security and Nutrition in the World 2023," Food Agriculture Organization of the United Nations, 2023, https://doi.org/10.4060/cc3017en.

110 *Against this backdrop:* "Food Waste in America in 2025: Statistics & Facts," Recycle Track Systems, https://www.rts.com/resources/guides/food-waste-america/.

110 *Families in Yemen:* Mohammed Ali Tharner, "Poverty in Yemen: Tracing the Path to Economic Downturn," Sada, September 12, 2024, https://carnegieendowment.org/sada/2024/09/poverty-in-yemen-tracing-the-path-to-economic-downturn?lang=en.

110 *We now grow crops:* Marc Brazeau, "The Steep Climb of Vertical Farms and Where Urban Ag Might Be Revolutionary," iGrow Pre-Owned, June 13, 2019, https://www.igrowpreowned.com/igrownews/the-steep-climb-of-vertical-farms-and-where-urban-ag-might-be-revolutionary-1.

110 *By 2030, it's projected:* Market Research Future, "Vertical Farming Market Is Booming and Projected to Hit $33.5 Billion by 2032, at 20.62% CAGR," EIN Presswire, March 12, 2025, https://www.einpresswire.com/article/793122056/vertical-farming-market-is-booming-and-projected-to-hit-33-5-billion-by-2032-at-20-62-cagr.

110 *And by 2050, climate change:* United Nations Foundation, "Global Agricultural Yields Could Decline 30% by 2050," Giving Compass, September 2, 2020, https://givingcompass.org/article/climate-change-and-the-future-of-food.

112 *By 2022, 95 percent:* Emily A. Vogels, Risa Gelles-Watnick, and Navid Massarat, "Teens, Social Media and Technology 2022," Pew Research Center, August 10, 2022, https://www.pewresearch.org/internet/2022/08/10/teens-social-media-and-technology-2022/.

112 *and for those between the ages:* "Facts About Suicide," CDC, March 26, 2025, https://www.cdc.gov/suicide/facts/index.html.

112 *In the late 1990s:* Art Van Zee, "The Promotion and Marketing of OxyContin: Commercial Triumph, Public Health Tragedy," *American Journal of Public Health* 99, no. 2 (2009): 221–27, https://www.ncbi.nlm.nih.gov/pmc/articles/PMC2622774/.

112 *And fentanyl delivered:* "Drug Enforcement Administration Announces the Seizure of Over 379 Million Deadly Doses of Fentanyl in 2022," Drug Enforcement Administration, December 20, 2022, https://www.dea.gov/press-releases/2022/12/20/drug-enforcement-administration-announces-seizure-over-379-million-deadly.

113 *In 2019, the last female:* Agence France-Presse in Shanghai, "One of Last Four Giant Softshell Turtles Dies in Chinese Zoo," *The Guardian*, April 15, 2019, https://www.theguardian.com/world/2019/apr/15/last-female-world-rarest-yangtze-giant-softshell-turtle-species-dies-chinese-zoo.

114 *Since 1970, wildlife populations:* "69% Average Decline in Wildlife Populations Since 1970, Says New WWF Report," World Wildlife Fund, October 13, 2022, https://www.worldwildlife.org/press-releases/69-average-decline-in-wildlife-populations-since-1970-says-new-wwf-report.

114 *A 2019 United Nations report found:* "UN Report: Nature's Dangerous Decline 'Unprecedented'; Species Extinction Rates 'Accelerating,'" United Nations Sustainable Development Goals, May 6, 2019, https://www.un.org/sustainabledevelopment/blog/2019/05/nature-decline-unprecedented-report/.

114 *Global fertility rates dropped:* Max Roser, "The Global Decline of the Fertility Rate," Our World in Data, 2014, https://ourworldindata.org/global-decline-fertility-rate.

115 *Humans use between 25 and 44 percent:* "Human Consumption of Net Primary Production," Earth Observatory, June 25, 2004, https://earthobservatory.nasa.gov/images/4600/human-consumption-of-net-primary-production.

115 *As environmental journalist Richard Manning once wrote:* Richard Manning, "'The Oil We Eat' Following the Food Chain Back to Iraq," Resilience, May 23, 2004, https://www.resilience.org/stories/2004-05-23/oil-we-eat-following-food-chain-back-iraq/.

115 *We've also co-opted 77 percent:* Hannah Ritchie, "50% of All Land in the World Is Used to Produce Food," World Economic Forum, December 11, 2019, https://www.weforum.org/stories/2019/12/agriculture-habitable-land/.

115 *And the result is:* Hannah Ritchie, "Wild Mammals Make Up Only a Few Percent of the World's Mammal Biomass," Our World in Data, December 15, 2022, https://ourworldindata.org/wild-mammals-birds-biomass.

116 *DeepMind's artificial intelligence:* Linda Geddes, "DeepMind Uncovers Structure of 200m Proteins in Scientific Leap Forward," *The Guardian*, July 28, 2022, https://www.theguardian.com/technology/2022/jul/28/deepmind-uncovers-structure-of-200m-proteins-in-scientific-leap-forward.

116 *A 2023 Goldman Sachs report:* "Humanoid Robots: Sooner Than You Might Think," Goldman Sachs, November 15, 2022, https://www.goldmansachs.com/insights/articles/humanoid-robots.

117 *Today, the poorest half of humanity:* "Global Economic Inequality: Insights," World Inequality Report 2022, October 20, 2021, https://wir2022.wid.world/chapter-1/.

120 *With a 30 percent fatality rate:* "The Triumph of Science: The Incredible Story of Smallpox Eradication," National Foundation for Infectious Diseases, May 8, 2023, https://www.nfid.org/the-triumph-of-science-the-incredible-story-of-smallpox-eradication/.

120 *Since 2009, scientists have identified:* Aliza Chasan, "Prepare for Next Pandemic, Future Pathogens with 'Even Deadlier Potential' Than COVID, WHO Chief Warns," CBS News, May 23, 2023, https://www.cbsnews.com/news/next-pandemic-threat-pathogen-deadlier-than-covid-world-health-organization/.

120 *The Center for Global Development:* Eleni Smitham and Amanda Glassman, "The Next Pandemic Could Come Soon and Be Deadlier," Center for Global Development, August 25, 2021, https://www.cgdev.org/blog/the-next-pandemic-could-come-soon-and-be-deadlier.

120 *DNA sequencing has dropped:* Nina Komadina, "The Evolution of DNA Sequencing Costs: Insights from 2001 to 2022," DataHub, February 4, 2025,

https://datahub.io/blog/the-evolution-of-dna-sequencing-costs-insights-from-2001-to-2022.

120 *The 5 percent of the population:* Ana Sanz-García, Clara Gesteira, Jesús Sanz, and María Paz García-Vera, "Prevalence of Psychopathy in the General Adult Population: A Systematic Review and Meta-Analysis," *Frontiers in Psychology* 12 (August 45, 2021), https://doi.org/10.3389/fpsyg.2021.661044.

120 *After five people died:* Scott Shane, "F.B.I., Laying Out Evidence, Closes Anthrax Case," *New York Times*, February 19, 2010, https://www.nytimes.com/2010/02/20/us/20anthrax.html.

120 *The CDC estimates that another:* Arnold F. Kaufmann, Martin I. Meltzer, and George P. Schmid, "The Economic Impact of a Bioterrorist Attack: Are Prevention and Postattack Intervention Programs Justifiable?," *Emerging Infectious Diseases* 3, no. 2 (1997): 83–94, https://doi.org/10.3201/eid0302.970201.

121 *Hinton left his job at Google:* Cade Metz, "'The Godfather of A.I.' Leaves Google and Warns of Danger Ahead," *New York Times*, May 1, 2023, https://www.nytimes.com/2023/05/01/technology/ai-google-chatbot-engineer-quits-hinton.html.

121 *A few weeks later, Hinton:* "Statement on AI Risk," Center for AI Safety, https://aistatement.com.

122 *A system that might replace:* Orianna Rosa Royle, "Silicon Valley Billionaire Vinod Khosla Says AI Will Handle 80% of Work in 80% of Jobs," *Fortune*, September 24, 2024, https://fortune.com/2024/09/24/silicon-valley-billionaire-vinod-khosla-universal-basic-income-ai-80-jobs/.

122 *Consider BMW's experiment:* Will Knight, "How Human-Robot Teamwork Will Upend Manufacturing," *MIT Technology Review*, September 16, 2014, https://www.technologyreview.com/2014/09/16/171369/how-human-robot-teamwork-will-upend-manufacturing/.

122 *In 1870, 70–80 percent:* Stanley Lebergott, "Labor Force and Employment, 1800–1960," in Dorothy S. Brady, ed., *Output, Employment, and Productivity in the United States After 1800* (National Bureau of Economic Research, 1966), https://www.nber.org/system/files/chapters/c1567/c1567.pdf.

122 *Today, it's 1.3 percent:* Isaac Arnsdorf, "How a Top Chicken Company Cut Off Black Farmers, One by One," ProPublica, June 26, 2019, https://www.propublica.org/article/how-a-top-chicken-company-cut-off-black-farmers-one-by-one.

122 *For every job:* "Internet Creates 2.4 Jobs for Every Job It Destroys: McKinsey," *Economic Times*, May 26, 2011, https://economictimes.indiatimes.com/tech/internet/internet-creates-2-4-jobs-for-every-job-it-destroys-mckinsey/articleshow/8586070.cms?from=mdr.

122 *We highlighted some of these:* Fabio Urbina, Filippa Lentzos, Cédric Invernizzi, and Sean Ekins, "Dual Use of Artificial Intelligence–Powered Drug Discovery,"

Nature Machine Intelligence 4 (2022): 189–91, https://doi.org/10.1038/s42256-022-00465-9.

123 *In 2024, AI-generated robocalls:* Ali Swenson and Will Weissert, "AI Robocalls Impersonate President Biden in an Apparent Attempt to Suppress Votes in New Hampshire," PBS News, January 22, 2024, https://www.pbs.org/newshour/politics/ai-robocalls-impersonate-president-biden-in-an-apparent-attempt-to-suppress-votes-in-new-hampshire.

123 *In his 1872 novel* Erewhon*:* Alissa Simon, "Revisiting Samuel Butler's Mechanical Kingdom in the Age of Artificial Intelligence," Harrison Middleton University, September 6, 2024, https://hmu.edu/revisiting-samuel-butlers-mechanical-kingdom-in-the-age-of-artificial-intelligence/.

123 *It's an intelligence explosion:* "'The Best or Worst Thing to Happen to Humanity'—Stephen Hawking Launches Centre for the Future of Intelligence," University of Cambridge, October 19, 2016, https://www.cam.ac.uk/research/news/the-best-or-worst-thing-to-happen-to-humanity-stephen-hawking-launches-centre-for-the-future-of.

124 *Eric Schmidt recently warned:* Paurush Omar, "Former Google CEO Eric Schmidt Sounds Alarm on AI Data Centers' Soaring Power Demand: 'We Need Energy in All Forms,'" *Economic Times*, May 12, 2025, https://economictimes.indiatimes.com/magazines/panache/former-google-ceo-eric-schmidt-sounds-alarm-on-ai-data-centers-soaring-power-demand-we-need-energy-in-all-forms/articleshow/121036712.cms?from=mdr.

124 *Microsoft just signed a contract:* C. Mandler, "Three Mile Island Nuclear Plant Will Reopen to Power Microsoft Data Centers," NPR, September 20, 2024, https://www.npr.org/2024/09/20/nx-s1-5120581/three-mile-island-nuclear-power-plant-microsoft-ai.

125 *Meanwhile, China added more:* Isabel Hilton, "How China Became the World's Leader on Renewable Energy," Yale Environment360, March 13, 2024, https://e360.yale.edu/features/china-renewable-energy.

PART 3

Chapter 7: Mind 2.0

129 *The brain is an energy hog:* Steven Kotler, *The Art of Impossible: A Peak Performance Primer* (Harper Wave, 2021).

130 *The brain is a prediction engine:* Karl Friston, "The Free-Energy Principle: A Unified Brain Theory?," *Nature Reviews Neuroscience* 11 (2010): 127–38, https://doi.org/10.1038/nrn2787.

130 *But in a world moving this fast:* Steven Kotler, Michael Mannino, Scott Kelso, and Richard Huske, "First Few Seconds for Flow: A Comprehensive

Proposal of the Neurobiology and Neurodynamics of State Onset," *Neuroscience & Biobehavioral Reviews* 143 (2022): 104956, https://doi.org/10.1016/j.neubiorev.2022.104956.

130 *Evolution optimized the human brain:* Kenneth Watson, review of *Thinking, Fast and Slow* by Daniel Kahneman, *Canadian Journal of Program Evaluation* 26, no. 2 (2011): 111–13, https://doi.org/10.3138/cjpe.26.010.

133 *The anterior cingulate cortex:* Kotler, *The Art of Impossible.*

133 *Our negativity bias amplifies threats:* Amos Tversky and Daniel Kahneman, "Prospect Theory: An Analysis of Decision Under Risk," *Econometrica* 47, no. 2 (1979): 263–92, https://www.jstor.org/stable/1914185.

133 *Status quo bias makes us resistant:* Roy F. Baumeister, Ellen Bratslavsky, Catrin Finkenauer, and Kathleen D. Vohs, "Bad Is Stronger Than Good," *Review of General Psychology* 5, no. 4 (2001): 323–70, https://doi.org/10.1037/1089-2680.5.4.323.

134 *A host of interventions:* Kotler, Mannino, Kelso, and Huskey, "First Few Seconds for Flow."

135 *In the brain, anxiety and excitement:* Gary Aston-Jones and Jonathan D. Cohen, "An Integrative Theory of Locus Coeruleus–Norepinephrine Function: Adaptive Gain and Optimal Performance," *Annual Review of Neuroscience* 28, no. 1 (2005): 403–50, https://doi.org/10.1146/annurev.neuro.28.061604.135709.

137 *In this creative state:* Kotler, *The Art of Impossible.*

137 *In the brains of creatives:* Kotler, *The Art of Impossible.*

137 *Additionally, the salience network:* Kotler, *The Art of Impossible.*

137 *Mihaly Csikszentmihalyi, the godfather:* Mihaly Csikszentmihalyi, *Creativity: Flow and the Psychology of Discovery and Invention* (HarperPerennial, 1996), p. 109.

138 *Gen Z'ers change jobs 134 percent faster:* Scott Galloway, "Gen Zers and Millennials Are Switching Jobs at an Accelerating Pace, and It's Paying off. Here's Where It Can Still Go Wrong," *Fortune*, April 23, 2024, https://fortune.com/2024/04/23/gen-zers-millennials-job-switching-accelerating-pros-cons-careers-employment/.

139 *Even fifteen minutes a day:* Jon Hamilton, "Building a Better Brain Through Music, Dance and Poetry," *All Things Considered*, NPR, April 3, 2023, https://www.npr.org/sections/health-shots/2023/04/03/1167494088/your-brain-on-art-music-dance-poetry.

139 *When Arie de Geus:* Arie de Geus, *The Living Company: Growth, Learning and Longevity in Business* (Harvard Business School Press, 1997).

139 *In May of 1961:* "What Was the Space Race?," National Air and Space Museum, Smithsonian Institution, August 23, 2023, https://airandspace.si.edu/stories/editorial/what-was-space-race.

140 *At the time, the United States had only:* Apollo 11 Press Kit (NASA, 1969), Apollo 11 Lunar Surface Journal, https://www.nasa.gov/history/alsj/a11/a11prskit.html.
140 *"We choose to go to the moon":* John F. Kennedy, Address at Rice University on the Nation's Space Effort, speech, Rice University, Houston, TX, September 12, 1962, https://www.rice.edu/jfk-speech.
140 *A clear mission:* Kotler, *The Art of Impossible.*
143 *Stanford psychologist Alia Crum:* Becca R. Levy, Martin D. Slade, Suzanne R. Kunkel, and Stanislav V. Kasl, "Longevity Increased by Positive Self-Perceptions of Aging," *Journal of Personality and Social Psychology* 83, no. 2 (2002): 262–70, https://doi.org/10.1037/0022-3514.83.2.261.
145 *Teller's insight was that it's often:* Astro Teller, "Google X Head on Moonshots: 10X Is Easier Than 10 Percent," *Wired*, February 11, 2013, https://www.wired.com/2013/02/moonshots-matter-heres-how-to-make-them-happen/.
147 *On January 15, 2009:* Amy Tikkanen, "US Airways Flight 1549," *Brittanica*, October 24, 2025, https://www.britannica.com/topic/US-Airways-Flight-1549-incident.
147 *Ninety seconds later:* Tikkanen, "US Airways Flight 1549."
148 *In flow, the prefrontal cortex:* Kotler, Mannino, Kelso, and Huskey, "First Few Seconds for Flow."
150 *But in a University of Sydney study:* Richard P. Chi and Allan W. Snyder, "Facilitate Insight by Non-Invasive Brain Stimulation," *PLOS One* (February 2, 2011), https://doi.org/10.1371/journal.pone.0016655.
150 *In 1899, James:* William James, *The Principles of Psychology* (Henry Holt, 1890).
151 *Flow states have triggers:* Kotler, *The Art of Impossible*; Kotler, Mannino, Kelso, and Huskey, "First Few Seconds for Flow."
152 *The brain has a 90–110 minute:* Tikkanen, "US Airways Flight 1549."

Chapter 8: The Androids Are Us

155 *The cognitive skills encountered:* P. N. Tandon, "The Decade of the Brain: A Brief Review," *Neurology India* 48, no. 3 (2000): 199–207, https://journals.lww.com/neur/fulltext/2000/48030/the_decade_of_the_brain___a_brief_review.1.aspx.
156 *Working memory taps out:* Steven Kotler, *The Art of Impossible: A Peak Performance Primer* (Harper Wave, 2021).
157 *In 2012, the pair:* Andrea Lo, "A Former Buddhist Monk Is Trying to Help the World Relax with an App," CNN, September 28, 2018, https://www.cnn.com/2018/09/28/business/headspace-meditation-app.
157 *Scientifically, 2005 was the inflection point:* Sara W. Lazar et al., "Meditation Experience Is Associated with Increased Cortical Thickness," *NeuroReport*

16, no. 17 (2005): 1893–97, https://doi.org/10.1097/01.wnr.0000186598.66243.19.

158 *Recovery scores correlate with productivity:* J. F. Thayer, Fredrik Åhs, Mats Fredrikson, John J. Sollers III, and Tor D. Wager, "A Meta-Analysis of Heart Rate Variability and Neuroimaging Studies: Implications for Heart Rate Variability as a Marker of Stress and Health," *Neuroscience & Biobehavioral Reviews* 36, no. 2 (2012): 747–56, https://doi.org/10.1016/j.neubiorev.2011.11.009.

158 *In 2009, she co-founded InteraXon:* "Understand Your Brain Patterns with Ariel Garten, Founder of Muse," Ness Labs, October 21, 2021, https://nesslabs.com/muse-featured-tool.

159 *Within a few years, the company:* Andrew Ali Aghapour, "My Brain on Muse, the Tech Meditation Headset," *The Revealer*, May 10, 2022, https://therevealer.org/my-brain-on-muse-the-tech-meditation-headset/.

161 *A 2024 McKinsey study:* Alex Singla, Alexander Sukharevsky, Lareina Yee, Michael Chui, and Bryce Hall, "The State of AI: How Organizations Are Rewiring to Capture Value," QuantumBlack AI, McKinsey, March 12, 2025, https://www.mckinsey.com/capabilities/quantumblack/our-insights/the-state-of-ai.

163 *psychologist Keith Sawyer:* Keith Sawyer, *Group Genius: The Creative Power of Collaboration* (Basic Books, 2007; rev. ed. 2017).

164 *a landmark 2021 study:* Mohammad Shehata et al., "Team Flow Is a Unique Brain State Associated with Enhanced Information Integration and Interbrain Synchrony," *eNeuro* 8, no. 5 (2021), https://doi.org/10.1523/ENEURO.0133-21.2021.

164 *to co-found Syneurgy:* "About Us," Syneurgy, https://www.syneurgy.com/about-us/.

165 *In a pilot study:* Michael Mannino and Erwin Valencia, "Syneurgy Internal Pilot Study Results" (unpublished company data, 2025).

165 *In 2005, Ray Kurzweil made one:* Ray Kurzweil, *The Singularity Is Near: When Humans Transcend Biology* (Viking, 2005).

166 *Steven was in the room:* "Computer Helps Blind Man 'See,'" *Wired*, January 17, 2000, https://www.wired.com/2000/01/computer-helps-blind-man-see/.

166 *The most famous example:* Tanya Lewis and Gary Stix, "Elon Musk's Secretive Brain Tech Company Debuts a Sophisticated Neural Implant," *Scientific American*, July 17, 2019, https://www.scientificamerican.com/article/elon-musks-secretive-brain-tech-company-debuts-a-sophisticated-neural-implant1/.

166 *Or, as Musk said:* Lex Fridman, "Elon Musk: Neuralink and the Future of Humanity," podcast, episode 438, August 2, 2024, https://lexfridman.com/elon-musk-and-neuralink-team/; transcript at https://lexfridman.com/elon-musk-and-neuralink-team-transcript/.

166 *than traditional stiff metal arrays:* Bill Chappell, "What to Know about Elon Musk's Neuralink, Which Put an Implant into a Human Brain," NPR,

January 30, 2024, https://www.npr.org/2024/01/30/1227850900/elon-musk-neuralink-implant-clinical-trial.

167 *In 2024, Neuralink implanted:* Michael Kan, "Neuralink Patient Also Uses Brain Chip to Play Mario Kart," *PCMag*, March 26, 2024, https://www.pcmag.com/news/neuralink-patient-also-uses-brain-chip-to-play-mario-kart.

167 *Unfortunately, the bandwidth:* Daniel Levitin, "Why It's So Hard to Pay Attention, Explained by Science," *Fast Company*, September 23, 2015, https://www.fastcompany.com/3051417/why-its-so-hard-to-pay-attention-explained-by-science.

167 *In experiments:* Jennifer Brown et al., "Optogenetic Stimulation of a Cortical Biohybrid Implant Guides Goal Directed Behavior," bioRxiv, November 22, 2024, https://doi.org/10.1101/2024.11.22.624907.

169 *Known as noninvasive neural decoding:* Jarod Levy et al., "Brain-to-Text Decoding: A Non-Invasive Approach via Typing," Meta AI Research, February 6, 2025, https://ai.meta.com/research/publications/brain-to-text-decoding-a-non-invasive-approach-via-typing/; "Using AI to Decode Language from the Brain and Advance Our Understanding of Human Communication," Meta AI, February 7, 2025, https://ai.meta.com/blog/brain-ai-research-human-communication/.

169 *LLMs trained:* Ziyi Ye et al., "Generative Language Reconstruction from Brain Recordings," *Communications Biology* 8, 346 (2025), https://doi.org/10.1038/s42003-025-07731-7.

172 *During meditation, the monks' brains:* Antoine Lutz, Lawrence L. Greischar, Nancy B. Rawlings, Matthieu Ricard, and Richard J. Davidson, "Long-Term Meditators Self-Induce High-Amplitude Gamma Synchrony During Mental Practice," *Proceedings of the National Academy of Sciences* 101, no. 46 (2004): 16369–73, https://doi.org/10.1073/pnas.0407401101.

173 *Additional studies found:* Judson A. Brewer, Patrick Worhunsky, Jeremy R. Gray, Yi-Yuan Tang, Jochen Weber, and Hedy Kober, "Meditation Experience Is Associated with Differences in Default Mode Network Activity and Connectivity," *Proceedings of the National Academy of Sciences* 108, no. 50 (2011): 20254–59, https://doi.org/10.1073/pnas.1112029108.

173 *During compassion meditation:* Tania Singer and Olga M. Klimecki, "Empathy and Compassion," *Current Biology* 24, no. 18 (2014): R875–78, https://doi.org/10.1016/j.cub.2014.06.054.

175 *Within days of the genome:* Fan Wu et al., "A New Coronavirus Associated with Human Respiratory Disease in China," *Nature* 579 (2020): 265–69, https://doi.org/10.1038/s41586-020-2008-3.

175 *The result was the fastest:* Florian Krammer, "SARS-CoV-2 Vaccines in Development," *Nature* 586, (2020): 516–27, https://doi.org/10.1038/s41586-020-2798-3; Gregory A. Poland, Inna G. Ovsyannikova, and Richard B. Kennedy, "SARS-CoV-2 Immunity: Review and Applications to Phase 3 Vaccine

Candidates," *The Lancet* 396, no. 10262 (2020): 1595–606, https://doi.org/10.1016/s0140-6736(20)32137-1.

Chapter 9: The Paradise Paradox

178 *The year is 1968:* John B. Calhoun, "Death Squared: The Explosive Growth and Demise of a Mouse Population," *Proceedings of the Royal Society of Medicine* 66, no. 1P2 (1973): 80–88, https://doi.org/10.1177/00359157730661P202.

179 *By day 315:* Calhoun.

179 *By day 600:* Calhoun.

180 *The last known birth:* Calhoun.

180 *In 1972, Calhoun published:* John B. Calhoun, "Population Density and Social Pathology," *Scientific American* 206, no. 2 (1962): 139–48.

181 *In papers and lectures:* Sam Kean, "The Disappearing Spoon: Death Squared," Science History Institute Museum and Library, November 29, 2022, https://www.sciencehistory.org/stories/disappearing-pod/death-squared/.

184 *Through this work:* J. Panksepp, *Affective Neuroscience: The Foundations of Human and Animal Emotions* (Oxford University Press, 1998).

184 *Panksepp's work overturned decades:* Panksepp.

185 *Panksepp found that play:* Jaak Panksepp and Jeff Burgdorf, "'Laughing' Rats and the Evolutionary Antecedents of Human Joy?," *Physiology & Behavior* 79, no. 3 (2003): 533–47, https://doi.org/10.1016/S0031-9384(03)00159-8.

185 *In our early years, billions:* Bryan Kolb and Robbin Gibb, "Brain Plasticity and Behaviour in the Developing Brain," *Journal of the Canadian Academy of Child and Adolescent Psychiatry* 20, no. 4 (2011): 265–76, PMC3222570.

187 *What emerged was the free energy principle:* Karl Friston, "The Free-Energy Principle: A Unified Brain Theory?," *Nature Reviews Neuroscience* 11, no. 2 (2010): 127–38, https://doi.org/10.1038/nrn2787.

187 *Children with a "growth mindset":* Jason S. Moser, Hans S. Schroder, Carrie Heeter, Tim P. Moran, and Yu-Hao Lee "Mind Your Errors: Evidence for a Neural Mechanism Linking Growth Mind-set to Adaptive Posterror Adjustments," *Psychological Science* 22, no. 12 (2011): 1484–89, https://doi.org/10.1177/0956797611419520.

188 *Once we started to be able:* Betsy Sparrow, Jenny Liu, and Daniel M. Wegner, "Google Effects on Memory: Cognitive Consequences of Having Information at Our Fingertips," *Science* 333, no. 6043 (2011): 776–78, https://doi.org/10.1126/science.1207745.

191 *Harvard geneticist:* David A. Sinclair with Matthew D. LaPlante, *Lifespan: Why We Age—and Why We Don't Have To* (Atria Books, 2019).

192 *In mice, they've:* James L. Kirkland, Tamara Tchkonia, Yi Zhu, Laura J. Niedernhofer, and Paul D. Robbins, "The Clinical Potential of Senolytic Drugs,"

Journal of the American Geriatrics Society 65, no. 10 (2017): 2297–301, https://doi.org/10.1111/jgs.14969.

192 *Google DeepMind:* John Jumper et al., "Highly Accurate Protein Structure Prediction with AlphaFold," *Nature* 596, no. 7873 (2021): 583–89, https://doi.org/10.1038/s41586-021-03819-2.

195 *One of the oldest rules:* Robert M. Yerkes and John D. Dodson, "The Relation of Strength of Stimulus to Rapidity of Habit-Formation," *Journal of Comparative Neurology and Psychology* 18, no. 5 (1908): 459–82.

195 *In the Harvard Study of Adult Development:* Robert Waldinger and Marc Schulz, *The Good Life: Lessons from the World's Longest Scientific Study of Happiness* (Simon & Schuster, 2023).

197 *In* A Theory of Fun*:* Raph Koster, *A Theory of Fun for Game Design*, 2nd ed. (O'Reilly Media, 2013).

ABOUT THE AUTHORS

STEVEN KOTLER is a *New York Times* bestselling author, award-winning journalist, and founder and executive director of the Flow Research Collective and the Flow Institute. He is the author of sixteen books, including twelve bestsellers. His work has been nominated for three Pulitzer Prizes and has appeared in over one hundred publications, including *The New York Times Magazine, Wired, The Atlantic, Time,* and *Harvard Business Review.*

Recognized as a pioneer in applied performance neuroscience, Kotler publishes peer-reviewed research on flow, altered states, creativity, intuition, and consciousness. The *New York Times* called him "one of the world's leading experts in ultimate human performance." His training programs have reached people in 156 countries and over 28 industries—including U.S. Navy SEALs; Olympic athletes; and executive teams at Google, Meta, Microsoft, Audi, and Accenture.

What Is Flow?

Flow is an optimal state of consciousness where we feel our best and perform our best. Attention deepens. Time slows. Self-talk disappears. Action and awareness merge and work that once felt effortful becomes effortless. It's the science behind "being in the zone."

What the Research Shows
McKinsey & Co: *up to 500 percent increase in productivity*
DARPA: *230 percent faster learning*
Harvard: *roughly 2 days of heightened creativity post-flow*
University of Western Australia: *80 percent improvement in performance*
Flow Research Collective: *8 weeks of training produces a 71.58 percent increase in time in flow.*

How to Get Involved
Today, Steven focuses his efforts on two organizations:

Flow Research Collective: A neuroscience-driven research organization advancing the science of flow and peak performance. FlowResearchCollective.com

Flow Institute: The training and development sister arm of the Flow Research Collective, the institute translates cutting-edge research into practical tools and workshops to help individuals and teams increase performance, productivity, and creativity. FlowInstitute.com

Steven also writes a weekly newsletter followed by five hundred thousand peak-performance enthusiasts, sharing insights from the front lines of flow research and performance neuroscience. He's also the host of *Flow Radio*, a top-ten Apple iTunes science podcast that's regularly featured on best-of lists across the web.

To learn more about Steven's trainings or to join hundreds of thousands receiving weekly downloads on flow, creativity, and consciousness, and also practical tips on performance, resilience, and recovery, please scan the QR code. qr.stevenkotler.com/wag

PETER H. DIAMANDIS is a *New York Times* bestselling author and founder of over twenty-five companies in the areas of AI, health tech, space, venture capital, and education. He is a cofounder of Singularity University and curator of Abundance360. He serves as cofounder of two $500 million+ venture funds: BOLD Capital Partners, focused on health and longevity; and Link-XPV, focused on AI. He is the founder and executive chairman of XPRIZE, which has launched $550 million of incentive prizes driving $10 billion in R & D. In the field of longevity, Dr. Diamandis is cofounder and chairman of Fountain Life. He has degrees in molecular genetics and aerospace engineering from MIT and an MD from Harvard Medical School. He has been named by *Fortune* magazine as one of the World's 50 Greatest Leaders. His *Moonshots* podcast focuses on AI and other exponential technologies, routinely exceeding one million listeners per episode.

To Get More Involved in Peter Diamandis's Ecosystem . . .

ABUNDANCE MEMBERSHIP AND ABUNDANCE SUMMIT

Abundance Membership

The Abundance Membership is six hundred leaders strong who have one shared mission: *using exponential technologies to transform success into global significance.* This is where leaders gather to go big, create wealth, and uplift humanity.

You've proven you can create value. Today, you're motivated to deploy your capital, network, and expertise toward solving challenges that matter at civilizational scale. We're living through the most consequential technological inflection point in human history. Artificial general intelligence is effectively here. Artificial superintelligence arrives within years. Historically, successful leaders optimize existing companies: better margins, incremental growth. But linear thinking will fail during the superexponential times ahead. Exponential leaders focus on moonshots. They pursue 10x growth, not 10 percent. Same resources. Exponentially different outcomes.

The Abundance Membership exists for leaders transitioning from "Success to Significance." This community of 600+ founders and entrepreneurs, leading $10 million to $10 billion companies, engineers an extraordinary future together. We believe the world's biggest problems are the world's biggest business opportunities. You have achieved success. The question now is significance.
Learn more at www.abundance360.com.

Abundance Summit

While others read about exponential technologies, those who attend the Abundance Summit learn directly from the innovators reshaping reality. In a Los Angeles coastal resort, the four-and-half-day summit focuses on AI, robotics, crypto, longevity, and moonshots. This isn't just a summit, it serves as an annual recalibration from linear thinking to exponential transformation, from scarcity viewpoints to abundance thinking.

Experience visionary conversations and keynotes from legendary founders and CEOs, as well as presentations from our exponential and Moonshot faculty, who have designed billion-person impact ventures. Play in our Tech Hub and Longevity Lounge, where hands-on participation allows you to feel and experience the coming exponential revolution. Past faculty includes Elon Musk, Sam Altman, Tony Robbins, Cathie Wood, Ray Kurzweil, Jacqueline Novogratz, David Sinclair, Palmer Luckey, Vinod Khosla, and more. The next ten years will define the next thousand. Join in person or virtually.
Learn more at www.abundance360.com/summit.

Peter's Metatrends Newsletter

DISCOVER THE FUTURE 10+ YEARS BEFORE EVERYONE ELSE.

Every week, Peter Diamandis and his team study the top 10 technology metatrends that will transform industries over the decade ahead: humanoid robotics, AI, quantum computing, transportation, energy, longevity, and more. No fluff. Only what matters. This isn't news, it's foresight. The signals everyone else will see in a decade, you'll see today. Read by founders and CEOs from the world's most disruptive companies; entrepreneurs who are building breakthrough technologies; leaders who refuse to be caught flat-footed by exponential change—those who build the future, not those who react to it.

It's a short, two-minute read delivered to your inbox every week: the most important trends impacting your life, your company, and your career—distilled into actionable intelligence. Gain access to trends ten years before anyone else. This is for you if you want to know what's coming, why it matters, and how you can benefit from it.
Subscribe at www.diamandis.com/metatrends.

Peter's *Moonshots* Podcast

If you enjoyed this book, I welcome you to continue the conversation by watching or listening to my podcast, *Moonshots*. Every week over one million people tune in to *Moonshots* and our special "WTF Just Happened in Tech" segment that delivers a blow-by-blow summary and interpretation of the hyperexponential world of AI, data centers, robotics, and crypto. This is where I track the future of technology and its impact on humanity in real time. I've had founders, investors, and technologists like Elon Musk, Cathie Wood, Ray Kurzweil, and more on the podcast to discuss how technology can create a better world for everyone. As you now know, we're living through a period of acceleration where technology changes overnight.

Moonshots is my effort to ensure my audience stays up to speed on everything that may impact their lives, careers, and businesses. Visit www.youtube.com/@peterdiamandis or scan the code to start listening to *Moonshots* today.